INFRARED THERMOGRAPHY IN THE EVALUATION OF AEROSPACE COMPOSITE MATERIALS

Related title

Anti-Abrasive Nanocoatings
(ISBN 978-0-85709-211-3)

Woodhead Publishing in Materials

INFRARED THERMOGRAPHY IN THE EVALUATION OF AEROSPACE COMPOSITE MATERIALS

Infrared Thermography to Composites

CAROSENA MEOLA, SIMONE BOCCARDI
AND GIOVANNI MARIA CARLOMAGNO

ELSEVIER

Amsterdam • Boston • Cambridge • Heidelberg
London • New York • Oxford • Paris • San Diego
San Francisco • Singapore • Sydney • Tokyo
Woodhead Publishing is an imprint of Elsevier

WP
WOODHEAD
PUBLISHING

Woodhead Publishing is an imprint of Elsevier
The Officers' Mess Business Centre, Royston Road, Duxford, CB22 4QH, UK
50 Hampshire Street, 5th Floor, Cambridge, MA 02139, USA
The Boulevard, Langford Lane, Kidlington, OX5 1GB, UK

British Library Cataloguing-in-Publication Data
A catalogue record for this book is available from the British Library

Library of Congress Cataloging-in-Publication Data
A catalog record for this book is available from the Library of Congress

ISBN: 978-1-78242-171-9 (print)
ISBN: 978-1-78242-172-6 (online)

For information on all Woodhead Publishing publications visit our website at https://www.elsevier.com/

Publisher: Matthew Deans
Acquisition Editor: Glyn Jones
Editorial Project Manager: Harriet Clayton
Production Project Manager: Debasish Ghosh
Designer: Maria Inês Cruz

Typeset by TNQ Books and Journals

The important thing is not to stop questioning. Curiosity has its own reason for existing. One cannot help but be in awe when he contemplates the mysteries of eternity, of life, of the marvellous structure of reality. It is enough if one tries merely to comprehend a little of this mystery every day. Never lose a holy curiosity.

Albert Einstein

CONTENTS

ABOUT THE AUTHORS

Carosena Meola is doctor in aeronautical engineering from the University of Naples (1981). Actually, she is a senior research staff member at the Department of Industrial Engineering, University of Naples Federico II. She is Department Consultant for Health and Safety at Work and Risk Management and a member of the Ergonomic Society (SIE). She has a Level III in infrared thermography according to both civil engineering (RINA) and aeronautical (Italian Aerospace Nondestructive Testing Board) standards and is a licenced instructor for personnel training and certification. Dr Meola's many credits include: member of UNI, CEN and ISO Technical Committees; member of the editorial board of ISRN Aerospace Engineering Journal, American Journal of Materials Science and Technology and Journal of Imaging; member of the Scientific Committee of AITA Conference; member of the Technical Program Committee of AMRA and SEIA conferences; chair of sessions within international conferences (AITA 2011, QIRT 2012, ECNDT 2014, SPB 2015); author of some 170 papers in well-recognized journals, books and proceedings and presenting author of several papers in many national and international conferences; editor of two books (about NDT and infrared thermography); and referee of over 40 international journals.

Simone Boccardi, an aerospace engineer, graduated from the University of Naples Federico II in 2013. He is currently a PhD student in the Department of Industrial Engineering of the University of Naples Federico II with his research topic being the application of infrared thermography to the investigation of composite materials. He is author of about 25 papers in well-recognized journals, books and proceedings and presenting author of several papers in national (AIVELA 2014) and international conferences (QIRT 2014, MetroAeroSpace 2015, ICCS 2015, AITA 2015).

Giovanni Maria Carlomagno is doctor in mechanical engineering, with honours, from the University of Naples (1965). His positions and accomplishments have included: assistant professor at the University of Naples (1967); assistantship in research at Princeton University, USA (1967/68); associate professor of physics and of gas dynamics (1969/1985), University of Naples; board of trustees member, University of Naples (1982/90); professor of aerospace and mechanical engineering, University of Naples (from 1986); dean of the Mechanical Engineering School (1989/92); member and chairman of von Karman Institute for Fluid Dynamics Technical Advisory Committee (1992/2015); member of the Academia Pontaniana (1992/now); member of the executive committee of the International Council of Aeronautical Sciences (1996/2008); Eminent Scientist Medal of the Wessex Institute of Technology, Southampton, UK (1999); fellow of the Japan Society for the Promotion of Science (1999); dean of the Aerospace Engineering School (1999/2006); Leonardo da Vinci Award for Flow Visualization (2000); editor or member of the editorial board of some 15 international scientific journals; author of some 400 scientific papers, of several books and of more than 10 industrial patents; editor or coeditor of some 30 books; FLUCOME Award (2007); Journal of Visualization Award (2007); president of the ARPA Research Consortium (2008/2012); honorary fellow of the International Council of Aeronautical Sciences (2008); and honorary member of the Quantitative InfraRed Thermography Council (2014).

ACKNOWLEDGEMENTS

We thank our students who carried out tests for their Master's thesis and contributed to investigate the effectiveness of infrared thermography, amongst them: Veronica Grasso, Francesca De Falco, Maria Esposito and Pasquale Ruocco.

A special thank goes to Giuseppe Sicardi for his technical support for testing rigs.

We want to thank all those who shared a stretch of the route contributing with specimens, testing apparatus and or financial support, amongst them: Profs Domenico Acierno, Giancarlo Caprino, Valentina Lopresto, Fabrizio Ricci, Michele Russo, Pietro Russo, Drs Natalino Daniele Boffa, Carmela Bonavolontà and Massimo Valentino.

CHAPTER 1

Composite Materials in the Aeronautical Industry

1.1 SOME HISTORICAL HINTS

From the first flight by the Wright brothers to aircraft becoming a common means of transport, materials have always been a key parameter in the evolution of aircraft, going from wood, to metal, to composites — searching for increasingly lightweight, high-temperature stability, and corrosion resistance features. In particular, weight/resistance ratio has always been a factor of great concern in aircraft technology.

At the beginning of the last century, during the pioneering phase, when aluminium was not yet available at reasonable prices, wood (a composite material provided by the natural world) was the only viable material to be used for the structures of a flying machine [1]. Indeed, at the time, wood was the cheapest and most readily available substance, easily tailored into the desired shape and strong enough to withstand flight loads. It was just with a wood-and-fabric biplane that the Wright brothers made the first flight on 17 December, 1903, achieving the first milestone in the aviation era.

A second important step in aeronautics was the so-called structural revolution of the 1930s, when wood was replaced by metal, mostly aluminium; such a revolution was marked by the Boeing 247D and the Douglas DC-3, even if, already in 1915, an all-metal construction was pioneered by Hugo Junkers, driven by military purposes.

Today most aircraft have their major parts made of composite materials. Composites were first introduced in military aviation in 1960 and, about 10 years later, also in the civil aviation. Initially, the use of composites was confined to the fabrication of secondary wing and tail components such as the rudder and wing trailing edge panels, involving directional reinforcement. A revolutionary exploitation of composites took place in the 2000s with the production of two big airplanes, the Airbus A380 and the Boeing Dreamliner; in fact, in both of these airplanes, composites have been extensively deployed in the primary load-carrying structure. They were followed by the A400, which is made almost entirely of composites.

Infrared Thermography in the Evaluation of Aerospace Composite Materials
ISBN 978-1-78242-171-9
http://dx.doi.org/10.1016/B978-1-78242-171-9.00001-2

The success of composites in the aviation field is mainly due to their favourable strength over weight ratio [2,3]; indeed, such a ratio is a chief parameter for an aerial craft that has to operate against the gravity force. Another reason lies in the fact that composites may be designed and tailored to fulfil many requirements. Conversely, the outcome may be a complex product that yet entails serious technical hitches in the complete understanding of its performance. The world of composites is an open endeavour that still needs to be fully explored!

1.2 BASICS OF COMPOSITES

A composite material is made of two or more basic substances that can be combined to obtain a new material, ie, the composite material, of enhanced properties with respect to its original constituents. The composites are generally made of a fibrous or particulate substance (*reinforce*) mixed within a *matrix* to form a relatively homogeneous material.

The matrix performs several critical functions, including maintaining the reinforcement-like fibres in the proper orientation and location, protecting them from abrasion and environmental effects, helping to transfer stresses among fibres, avoiding the propagation of fractures, and also contributing to electrical conductivity as well to thermal stability.

The materials that can be used as matrix include: cement, ceramics, metals and polymers.

Cement matrix composites are widely exploited in civil engineering, especially as concrete products in which sand, stones and steel act as reinforcement embedded in the form of particles or metal rods (reinforced concrete).

Ceramics and metals require very high temperatures and sometimes high pressures for processing; therefore, they are used mainly when a high-temperature strength and resistance to corrosion are compulsory.

Ceramic matrix composites (CMCs) are particularly appreciated for their resistance to exposure to high temperature and to environmental effects like corrosion; conversely, they are extremely brittle. Their primary use is for thermal protection systems (TPSs), for example, as carbon-reinforced silicon carbide (C/SiC) or silicon carbide reinforced silicon carbide (SiC/SiC), which are used where oxidation resistance and high-temperature capability are critical [4], especially for thermal protection in the aerospace field. Substantial advances have been made in the thermal barrier coating (TBC) of gas turbine blades and vanes.

Metal matrix composites (MMCs) are composed of a metal matrix and a reinforcement, or filler, material, which confers excellent mechanical performance [5]. Aluminium and its alloys are mostly used as matrix, but magnesium and titanium are employed too. Silicon carbide and graphite are generally chosen as reinforce. The main peculiarities of MMCs are: enhanced specific strength and stiffness, low density and high-temperature strength. In general, a magnesium matrix is used for manufacturing of parts for gearboxes, compressors and engines. Instead, a titanium matrix is mainly used for manufacturing of turbine engine components (fan blades, actuator pistons, synchronization rings, connecting links, shafts and discs). Moreover, in the aerospace field, Al-SiC, Al-B, Mg-C, Al-C, Al-Al_2O_3 continuous and discontinuous reinforcements are widely used for frames, reinforcements and aerial joining elements [6].

Due to their relative low-processing cost and weight, polymers represent the type of matrix most commonly used in non-civil products. Indeed, polymer matrix composites (PMCs) can be defined as market dominant among the other composites in the aircraft industry and will be treated in greater detail in this book. Reinforcements are often made of glass or carbon fibres.

1.3 POLYMERS

Roughly speaking, the term *polymer* indicates a large molecule constituted of a long chain of reiterating units (small molecules called monomers, or 'mers'), bonded together through a so-called polymerization chemical reaction. Polymerization requires at least two reaction points or functional groups for each monomer. There are two types of polymerization, condensation polymerization and addition polymerization. In the first, the chain development is accompanied by elimination of small molecules such as H_2O or CH_3OH; in the second, monomers react to form a polymer without formation of by-products, but, to get polymerization, the addition of catalysts is needed [7].

Owing to their behaviour under heating or cooling, polymers can be grouped into two categories: *thermosets* and *thermoplastics*.

Before polymerization, thermosets behave like low-viscosity resin, which cures gradually at a relative low temperature (20–200°C) and cannot be reprocessed by reheating. In a fully cured state, thermoset molecules are cross-linked and permanently insoluble and infusible. These types of polymers are also known as cross-linked polymers. Thermosets include unsaturated epoxies, polyesters and phenolics.

Cured epoxy resins are reasonably stable to chemical attack and are excellent adhesives having low shrinkage during curing (polymerization) and no emission of volatile gases. Because of these characteristics, which result in a material with high mechanical properties and high corrosion resistance (coupled with a quite simple curing process), they are the most popular amongst composite matrices. In contrast, epoxies are quite expensive, cannot be stored for a long time and so are mainly used in high-technology areas.

On the other hand, polyester resins are quite easily accessible, cheaper and used in a wide range of fields. Liquid polyesters can be stored at room temperature for months, sometimes even years, and with the mere addition of a catalyst can cure within a short time; the cured polyester can be rigid, or flexible, as the case may be, and transparent. They are mainly used in the automotive and naval fields.

Phenolics represent the first truly synthetic plastic (commercialized in 1905) obtained combining formaldehyde and phenol. They are water and solvent resistant, can be used as an electrical insulator (they were extensively used in circuit boards), but are generally brittle even if they can be strengthened, to a certain extent, by fillers. Nowadays, they have been practically superseded by modern plastics such as epoxy or polyester resins.

Thermoplastics are also linear polymers, which are composed of chain-like molecules, and may be high-viscosity resins with varying degrees of crystallinity; a number of them can be dissolved in certain liquids, and they soften or melt upon heating above their melting temperature (100–400°C) for additional processing. Some types of thermoplastic resins include polypropylene (PP), polyvinylchloride, polyether imide (PEI), polyether ether ketone (PEEK), polystyrene and polyphenylene sulfide (PPS). Thermoplastics have recently become of great interest for their ductility and high-processing speed as well as for the greater choice of manufacturing techniques. In fact, their processing can be selected by the scale and rate of production required and by the size of the component. In addition, thermoplastic composites can be easily repaired because transition to the softened phase can be accomplished any number of times by heating them up.

Due to the lightweight requirement, thermoset matrices (especially epoxy) are generally preferred for aeronautical applications, even if the use of the thermoplastic ones is emerging as the total composite percentage in the airplanes increases.

1.4 REINFORCEMENT FOR POLYMER-BASED COMPOSITES

The reinforcement provides most of the composite strength and stiffness. Different materials can be used as reinforcement and they can be introduced in the matrix in different formats such as particles, whiskers or fibres (Fig. 1.1). Particles are almost cubic pieces of material (Fig. 1.1(a)); whiskers are single stretched crystals (Fig. 1.1(b)); fibres are single filaments of very small diameter/length ratio (Fig. 1.1(c)), which can be aligned along one main direction (Fig. 1.1(d)) or randomly oriented (Fig. 1.1(e)).

Fibres are the most commonly used because a material with a fibrous shape entails small and scarce defects and fibres can be oriented along the main tensile stresses, with consequent high strength. In addition, fibres with small diameter offer great flexibility to fabrication processes. High-performance composites are generally manufactured with continuous fibres, while composites in less-demanding applications can be made with cheaper short fibres in an aligned array or with a random orientation.

1.4.1 Woven Fabric Architectures

The reinforcing medium can be also produced in the form of a woven fabric by directly interlacing either separate bundles of fibres, or tows, combining warp (0 degree) and weft (90 degree) in a regular pattern or

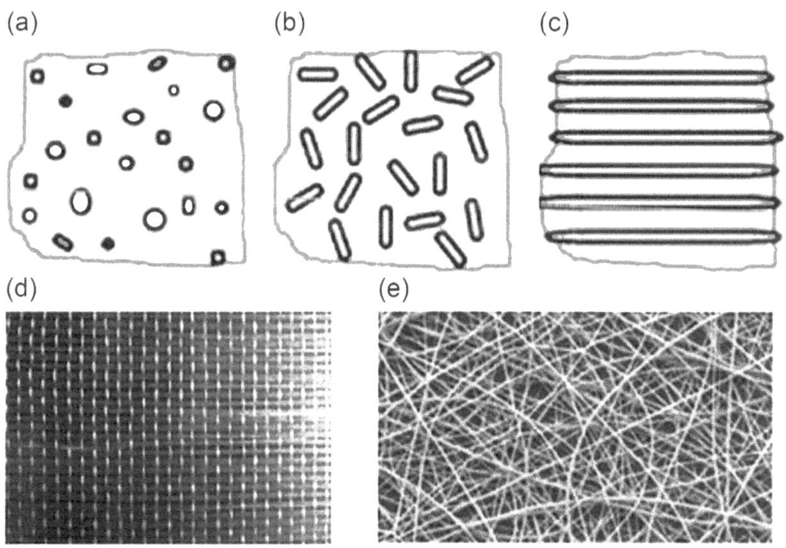

Figure 1.1 Reinforcement formats. (a) Particles, (b) whiskers, (c) fibres, (d) aligned fibres and (e) randomly oriented fibres.

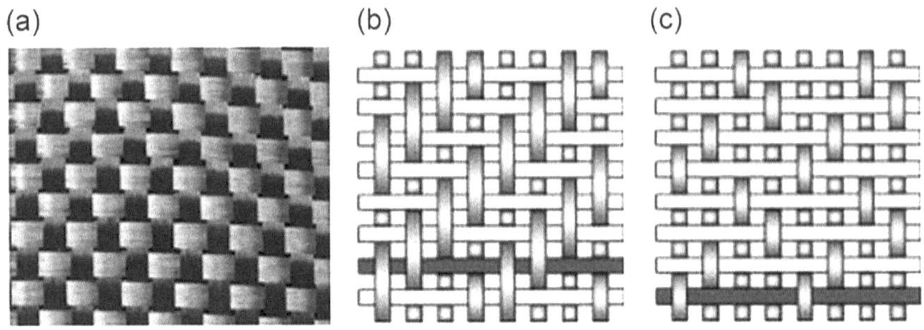

Figure 1.2 Some woven fabric types. (a) Plain, (b) twill and (c) satin.

weave style. The woven fabric architecture (Fig. 1.2) should be chosen with a compromise between ease of handling during manufacture, drapeability (the ability to form the fabric into a three-dimensional geometry) and mechanical performance [8].

For example, the compact plain weave (Fig. 1.2(a)), in which each warp fibre passes alternately under and over each weft fibre, is highly stable during handling, due to the intertwined weave structure, but it is the most difficult of the weaves to drape. In addition, it produces a composite with reduced in-plane strength and stiffness because of the high level of fibre crimp; the latter is a misalignment of fibres from the plane of the fabric, which produces resin-rich areas of limited performance benefits.

Superior wet out and drape is seen in the twill weave (Fig. 1.2(b)) where one or more warp fibres alternately weave over and under two or more weft fibres in a regular repeated manner with the visual effect of a straight or broken diagonal 'rib' to the fabric. Twill weaves also have reduced crimp, a smoother surface and slightly higher mechanical properties over the plain weave with only a small reduction in stability.

Satin weaves (Fig. 1.2(c)) are fundamentally twill weaves modified to produce fewer intersections of warp and weft. The 'harness' number used in their designation (typically 4, 5 and 8) is the total number of fibres crossed and passed under, before the fibre repeats the pattern. In particular, a 5-harness satin weave is one of the most drapeable examples, with a weave pattern of reduced intertwining, and produces improved in-plane mechanical properties at the expense of stable handling. In addition, the asymmetry needs to be considered in satin weaves since one face of the fabric has fibres running predominantly in the warp direction while the other face has fibres running predominantly in the weft direction.

1.4.2 Preimpregnated Thermosets

Either reinforcing unidirectional (UD) fibre tapes or woven fabrics can be pre-impregnated with a partially cured thermoset resin to produce what is known as a 'pre-preg'. These compounds form the basis of most high-performance structural composite components and have been used for many decades. The material needs storage at low temperature to prevent the resin from prematurely curing, which could start at room temperature. The material, supplied in rolls or tape, is cut into pieces and laid up in a mould manually or robotically, then, it must be vacuum bagged prior to the start of the curing process, which generally takes place in an *autoclave*, under a controlled pressure/temperature cycle.

The resulting structures made from pre-preg materials can show significant performance benefits compared to those made from alternative continuous fibre materials, with an increase of stiffness and strength up to 30% [9]. These advantages are to be ascribed to a high nominal fibre volume fraction of 60% and the use of high-grade, fully impregnated resins. Adversely, the production of such materials is expensive in terms of equipment required, storage and skilled labour costs.

1.4.3 Non-crimp Fabric Architectures

Non-crimp fabrics (NCFs) differ from woven fabrics by a stitching material (polyester yarn) that is introduced to bind a number of unidirectional fibre layers (Fig. 1.3) to avoid misalignments. The benefits of overlaying straight tows joined by stitching is that tow crimp may be avoided, which results in improved strength, stiffness and fatigue life. As with woven fabrics, a variety of assembly styles are available [10] involving different mass per unit area (defined by the number of unidirectional layers in the

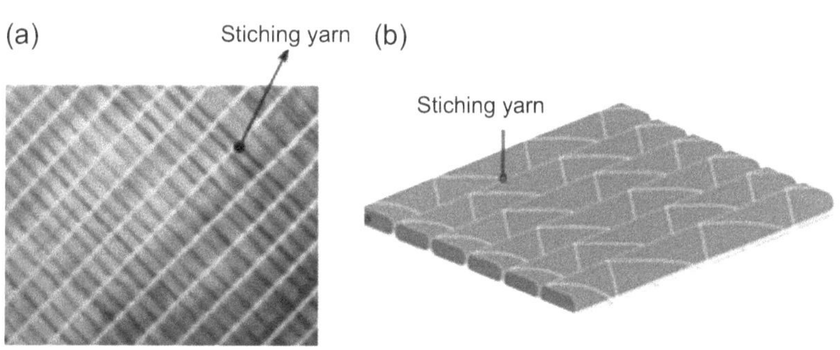

Figure 1.3 Example of stitching in non-crimp architecture. (a) Unidirectional and (b) biaxial.

fabric and the number of individual fibres contained in a single tow) and stitching mechanisms.

The manufacture is simple, using conventional stitching machinery, and the absence of tow crimp may result in composites of higher mechanical performance as compared to woven fabrics. The NCFs show some reduction in performance as compared to those derived from pre-pregs mainly because of the lower fibre volume fractions and the use of lower-performance resins suitable for infusion. The fibre over resin percentage may be enhanced by reducing the tow spacing and by improving the stitching architecture as well the overall manufacturing process, leading to a product of comparable performance but cheaper than those from pre-pregs.

1.4.4 Fibres Types

The most common fibre types used in composite industry are glass, carbon and organic (Kevlar). Other types, such as boron, silicon carbide and alumina are used in dedicated applications.

1.4.4.1 Glass Fibres

Fibreglass was first discovered in 1893 and made commercially available in 1936; it was first used as insulation material in electrical, thermal and acoustic uses. Then, it achieved a great popularity during the 1950s when it was considered as a good substitute to asbestos fibres whose health hazards were becoming apparent. Today fibreglass is the dominant reinforcement fibre in composite construction, accounting for about 90% of worldwide consumption. This is simply because it has good strength-to-weight characteristics, can be easily processed and sells at a low price.

Glass filaments are relatively easily produced by extruding molten glass, which is obtained by blending quarry products (sand, kaolin, limestone, colemanite) at about 1600°C; then, the formed liquid is passed through micro-fine bushings and simultaneously cooled to produce the fibre filaments of diameter generally ranging between 5 and 24 μm. The filaments are drawn together into a strand (closely associated) or roving (more loosely associated), and coated with a '*size*', or binder; this is to provide filament cohesion and to minimize degradation of filament strength that would otherwise be caused by filament-to-filament abrasion. The size may be temporary, as in the form of a starch-oil emulsion that is subsequently removed by heating and replaced with a glass-to-resin coupling agent known as a *finish*. On the other hand, the size may be a compatible treatment that performs several necessary functions during the subsequent forming operation and which, during impregnation, acts as a coupling agent to the resin being reinforced [11].

By varying the 'recipe' (ie, addition of chemicals to silica sand), different types of glass can be produced:

- A-glass (alkali glass) has good chemical resistance, but lower electrical properties.
- C-glass (chemical glass) has very high chemical resistance.
- E-glass (electrical glass) is an excellent insulator and resists attacks from water.
- R-, S- and T-glass (structural glass) are optimized for mechanical properties; the different letter identifies the manufacturer's trade name for equivalent fibres.
- D-glass (dielectric glass) has the best electrical properties but lacks in mechanical properties when compared to electrical and structural glass.
- M-glass (modulus) has high stiffness.

Electrical and structural glasses are, by far, the most common types found in composites because of their good combination of chemical resistance, mechanical properties and insulating properties. In particular, E-glass looks more attractive from the cost point of view, while structural glass offers better mechanical performance.

1.4.4.2 Carbon Fibres

Carbon has the highest strength and highest price of all reinforcement fibres today available for composites. These fibres were produced in the United Kingdom in the early 1960s, even if Edison had much earlier used them in lighting lamps. The most common method of making long carbon fibres is the oxidation and thermal pyrolysis of an organic precursor, poly-acrylonitrile (PAN). Through heating at correct conditions (2500−3000°C), the non-carbon constituents evaporate away with a resulting material having a 93−95% carbon content. Of course, the properties of carbon fibres depend on the raw material and the manufacturing process; in fact, the relative amount of exposure at high temperatures 500−3000°C results in greater or less graphitization of the fibre. Higher degrees of graphitization usually result in a stiffer fibre (higher modulus) with greater electrical and thermal conductivity values.

The size, or thickness, of carbon tows is measured in 'k' or thousands of filaments. A 3k tow contains 3000 filaments and a 12k has 12,000. Carbon fibres exhibit: substantially better strength and stiffness values than all the others types for fibre reinforcement, outstanding temperature performance, and high electrical and thermal conductivities. Impact or damage tolerance of pure carbon composite products can be from relatively low to very poor and greatly depends on the processing method. Despite that, when weight on a composite product is important, carbon fibres represent the best reinforcement to use because of the significant advantages retained by them:

- High stiffness-to-weight ratio
- High strength
- Corrosion resistant
- Fatigue resistant
- High-energy absorption on impact
- Tailored material properties

First of all, it has to be mentioned that carbon fibres are very light, resulting in lightweight structures. Furthermore, one can chose between stiff or strong fibres depending on the composite part being produced. Another major advantage is that their thermal expansion is basically zero; this means that unlike metals, which expand when heated, carbon fibres remain in their basic form with remarkable benefits in specific projects where thermal stability is required. Moreover, the material can resist very high temperatures (1000°C), being practically limited only by the matrix. If properly designed and conceived, carbon fibre composite structures do not suffer any fatigue issues. Finally, carbon fibres are permeable to X-ray and do not corrode, which is a huge concern with metals.

The material also has some disadvantages that need to be taken into consideration when planning a project. For instance, carbon fibres are fairly expensive compared to other reinforcements even if their price is steadily decreasing due to the progress of production technology. Moreover, carbon fibre is an electric conductor and, as such, can reflect radio waves, which can be a disadvantage in some cases. In addition, carbon fibres are brittle and material breakage can create debris, which can fly in multiple directions with safety implications.

Due to their intrinsic brittleness, handling of carbon fibres may be difficult and some protection is necessary. The size material must be appropriately chosen since it must provide consistent handling without swelling residues on the processing equipment and without obstacles to the penetration of resin into the fibre bundle. Owing to the different commercial carbon fibre surface features (ie, smooth, striated, round or kidney shaped), a different blend of physical characteristics optimized for the fibre shape and surface texture is required. Size materials must also be compatible with the resin matrix; this includes solubility in and/or reactivity with the formulated resin. This allows the resin to better penetrate the fibre bundle and interact with the fibre surface.

1.4.4.3 Aramid Fibres

The aramid fibres have high energy absorption during failure, which makes them ideal for impact and ballistic protection. Because of their low density, they offer high tensile strength-to-weight ratio, and high modulus-to-weight ratio, which makes them attractive for aircraft and body armour. Conversely, they have relatively poor shear and compression properties, which demand careful design for their use in structural applications that involve bending or compression.

Aramid was invented by DuPont in the 1960s (Kevlar is the registered trade name of DuPont aramid) as a result of research on nylon (polyamide) fibres. It was first introduced in the market in the 1970s as tire reinforcement and, like fibreglass, as a substitute for asbestos. The aramid chemical structure shows the aromatic benzene rings along the polymeric backbone; indeed, the word *aramid* derives from the blend of aromatic polyamide.

Aramid has good strength and modulus, with compression and shear strength similar to E glass, but low density and UV resistance. In addition, it has excellent toughness and resistance to impact, damage, abrasion and heat (up to 500°C), which makes aramid ideal for use in armour, military and ballistic applications, like helmets and bulletproof vests, as well for fire-fighting protection, on the underside of airplanes (protection against stone hits during takeoff and landing) and the underside of race cars. In general, any structure that is exposed to impact, abrasion and/or heat may benefit from the use of aramid. In this context, a good compromise may be the 'hybrid' carbon-aramid construction, which shares the high strength and stiffness of carbon and the impact protection of aramid.

However, processing aramid in composite fabrication is somewhat more difficult than fibreglass and carbon. In fact, there are some drawbacks:

- the high toughness makes fabrics difficult to cut with conventional tools
- wet out of fibres is more difficult than with fibreglass and carbon
- orthophthalic polyester resin will not adhere well to aramid
- postfabricating aramid components, eg, trimming off the edges, are also demanding

1.4.4.4 Boron Fibres

Boron fibres are produced by a chemical vapour deposition process. Boron can be deposited on a tungsten wire core, as well as on a glass or graphite filament core. Thus fibres of nominal diameter in the range 0.1–0.2 mm can be produced. They are characterized by low density, high tensile strength and high modulus of elasticity. In particular, boron fibres are extremely stiff (eg, five times stiffer than glass fibres) and difficult to weave, braid or twist, but they can be formed into resin-impregnated tapes for hand lay-up and filament winding processes. The high cost of boron filaments has greatly limited their use in experimental aircraft and space applications.

1.5 MANUFACTURING PROCEDURES

Composites can be produced through different manufacturing procedures, owing to different factors involving the type of reinforcement (pre-impregnated, or dry) and costs. For instance, in the past, the production of composites was mainly dominated by the pre-preg philosophy, with the primary challenge to achieve materials of high performance to compete with, or surpass, metals. Production with pre-pregs is expensive and, today, the attention is ever more toward new fabrics and alternative cheaper manufacturing processes. Within this context, composites derived from dry fabrics and liquid infusion processing are being considered for applications in various structural components instead of pre-preg-based composites. As an example, NCF composites, involving a multiaxial reinforcement (made up of multiple plies of dry fibres layered on top of each other and stitch bonded with a polyester thread), represent a good compromise amongst fibre strength and costs.

In the following sections some of the most important manufacturing processes are briefly described.

1.5.1 Hand Lay-up

The hand lay-up is still the mostly used manufacturing technique with pre-pregs. It is based on the superimposition of pre-impregnated laminas of thickness in the range $0.125 \div 0.30$ mm, (ie, laminas of unidirectional or woven fibres partially impregnated with resin). The raw material is generally found in rolls or tapes, from which laminas of given dimensions are cut and layered on top of each other, according to assigned shape and stacking sequence; the latter entails the orientation of fibres and the whole thickness. Once the laminas are stacked, a vacuum bag is applied and an autoclave cycle curing is performed. Generally, the curing cycle involves cooperation between temperature and vacuum to achieve two main purposes: reach the correct viscosity level of the resin for its uniform distribution through the laminate; eliminate the entrapped air, or other developed gases.

This procedure allows fabrication of materials with optimal characteristics, but it requires skilled personnel.

1.5.2 Filament Winding

This process requires tows of pre-impregnated fibres and a mandrel. The tows are continuously wound onto the mandrel following an ad hoc design by dedicated software in order to obtain a specific fibre orientation and quantity [12]. After winding, the wounded mandrel is put in a vacuum bag and then in an autoclave for curing. The procedure is completely automated and allows fabrication of pressure vessels, aircraft bodies, power poles, pipes and much more.

1.5.3 Pultrusion

This process was conceived in the 1950s by Goldsworthy who is considered the initiator of composites. It consists in pulling raw composites into a shaped die, including fibre impregnation pre-form and curing. More specifically, the pultrusion process starts with racks, or creels, holding rolls of fibre mat or doffs of fibre roving. The raw fibre is pulled off the racks and guided through a resin impregnation system (in some pultrusion systems, resin can also be injected directly into the die) and, after, guided through a series of tooling (pre-form), which organize the fibre into the correct shape, while excess resin is squeezed out. Finally, the obtained composite passes through a heated steel die where an exothermic reaction cures the resin matrix.

The component is continuously pulled and exits the mould as a hot, constant cross-sectional part that is generally cooled by forced air convection and then cut into the preferred dimension. This process is cost-effective for high volume production of constant cross-section parts.

1.5.4 Resin Transfer Moulding

Resin transfer moulding (RTM) [13] represents a viable alternative to prepregs and autoclave-based composites production. It includes a rigid closed mould. The process starts with the reinforcement mat, or woven roving, being draped in the bottom half of the mould. Then, the top mould is closed and catalyzed, low-viscosity (heated) resin is pumped in under pressure, displacing the air and venting it at the edges until the mould is filled. The mould is clamped and the resin allowed to cure. As main advantages of RTM, the composite dimensions depend only on the mould shape, the obtained composites have good surface finish on both of their sides and they are ready to be used without further treating. This process is cheaper than others and is gaining popularity; many aerospace components are manufactured by RTM, such as door pillars, stringers, stiffened panels, rudder tips and ribs.

1.5.5 Resin Film Infusion

Unlike RTM, the resin film infusion (RFI) process requires just one male, or female, mould of the desired shape [14]. Dry fabrics are laid up interleaved with layers of pre-catalyzed semisolid resin film supplied on a release paper. Then, the lay-up is vacuum bagged to remove air through the dry fabrics. Finally, the resin is thermally cured in an oven or an autoclave, while vacuum is applied. This process allows to reach high specific strength but it is not generally used for manufacturing of structural parts.

1.5.6 Resin Infusion

This process is similar to the RFI with the difference that only dry fibres are placed in the mould, which is closed in a vacuum bag. The one-sided mould is connected with a resin source and a vacuum pump. The liquid resin with hardener infuses into the reinforcing fibres thanks to the vacuum drawn through the mould. Curing and de-moulding steps follow the impregnation process to finish the product.

One main advantage of this process resides in the possibility to produce large and thin parts in a relatively inexpensive way.

1.6 FIBRE METAL LAMINATES

Fibre metal laminates (FMLs) combine the high bearing strength and impact resistance of metals with the excellent fatigue characteristics, high strength and stiffness, as well corrosion resistance of fibre-reinforced polymers, thus overcoming most of the disadvantages of single monolithic materials sheets. FMLs were conceived at the Delft University of Technology in the Netherlands, starting from the end of the 1970s, with the idea to increase the fatigue performance of aluminium alloys. An improvement was observed by introducing aramid fibres into the adhesive layers, so that the first FMLs were called aramid fibre-reinforced aluminium laminates (ARALL). Four different types of ARALL were investigated passing transitioning to carbon fibres (CARALL) and then to glass fibres (GLARE); for an historical overview, see Vlot and Gunnink, Fibre Metal Laminates: An Introduction [15].

The most successful FML is Glare, which seems to have the better impact resistance, as well mechanical properties after impact, than any other material. In particular, unlike classical composites, Glare undergoes plastic deformation under impact with the indenting of the external aluminium layer, which is easily noticed by the naked eye. In addition, the growth rate of fatigue cracks, which is normally high in aluminium sheets, is significantly reduced in the presence of glass fibres. As a most important feature, Glare has good burn-through resistance and so it is the optimal material for fabrication of vital parts, its main disadvantage being the cost. Today in modern aircraft, Glare is being mostly used for the construction of the fuselage.

1.7 SANDWICH STRUCTURES

To provide extra strength and shape, a lightweight core of a material is inserted between two sheets of laminates (skins) to form a sandwich structure. Various types of core and skins can be used depending on the performance requirements of the final part. Considering that a sandwich structure must be strong and lightweight at the same time, a good solution is to insert a cellular structure like a foam, or honeycomb (Fig. 1.4), between two thin composite laminates. For aeronautical applications, skins are generally made of carbon/epoxy, while the core may be Nomex, which is a honeycomb made from aramid paper.

Recently, a new class of materials, such as metal foams, are being considered as core in sandwich structures. The foam is generated by forced expansion of a precursor, which is generally an aluminium alloy mixed with a percentage of foaming agent. As a great advantage, a metal foam almost maintains the mechanical characteristics of the initial solid material, but it is

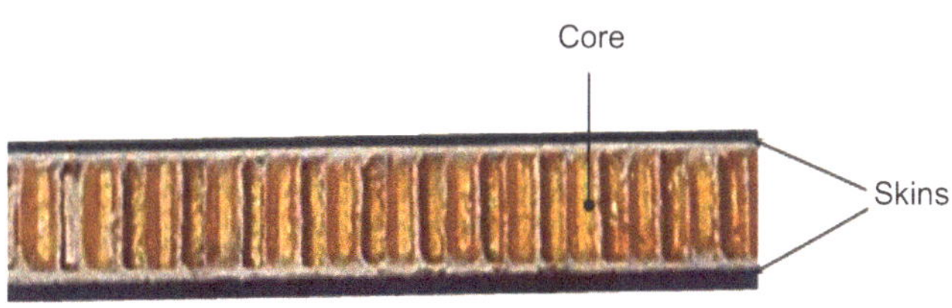

Figure 1.4 Example of sandwich structure.

much lighter. Therefore, it has good stiffness and strength-to-weight ratios as well as the ability to absorb impact energy and electromagnetic waves; then, thanks to the air pockets contained in its structure, a metal foam can also be used as acoustic and/or thermal insulator.

1.8 MAIN TYPES OF DEFICIENCY IN COMPOSITES

Being composites made of two or more basic materials and manufactured through complex processes involving temperature, pressure, chemical reaction etc., it may be expected that the final product can be affected by anomalies. Moreover, the in-service life of the components could be responsible for defect creation and/or degradation of the composite. In the following discussion, the defects that are mostly present in composites will be grouped into two categories: manufacturing defects and in-service failures.

1.8.1 Manufacturing Defects

Several different types of defects may occur during fabrication of composites, the most common being fibre/play misalignment, broken fibres, resin cracks or transversal ply cracks, voids, porosity, slag inclusions, nonuniform fibre/resin volume ratio, disbonded interlaminar regions, kissing bonds, incorrect cure and mechanical damage around machined holes and/or cuts.

The effective performance of a fibre is a function of its correct alignment with the stress/strain direction. In fact, in the presence of fibre misalignment, the loading of the fibre may change from straight tension/compression to shear loading of the weaker interface. This may result in a considerable drop of the composite mechanical properties [16].

Inclusions such as dirt and debris, which may be inadvertently present in composites, may contaminate the matrix or act as local stress concentrators, leading to delamination during the manufacturing process or when the composite is in service.

The strength of a carbon-fibre-reinforced polymer (CFRP) component is strongly dependent on the volume percentage of resin with respect to fibres. In fact, the presence of regions of fibres unsupported by the matrix can induce local stress concentration (notch effect), with a consequent severe degradation of strength and stiffness during the component's in-service life [17].

Amongst the procedures that allow composite production, the most important is probably the resin curing one. This process has to be optimized in order to get appropriate components responding to the structural design requirements. Otherwise, resin curing is dependent on the temperature rate increase, the temperature and duration of the curing plateau, the time at which pressure is applied and the post-curing temperature and pressure. If something unexpected occurs in the process, the typical consequence could be incomplete or inappropriate chemical reactions, uneven wetting of the fibres, incorrect fibre—volume ratio as well as formation of local matrix-rich pockets or matrix-starved regions. Moreover, the vacuum pressure, if not suitable, could affect the degassing of contaminants with only partial removal of the gases developing during chemical reactions. This may induce formation of voids (or porosity) within the matrix, between the plies or at the fibre/matrix interface. The detrimental effects of porosity have been known since 1978 when it was found that there was a decrease of the interlaminar shear strength by about 7% for each 1% of voids up to a total void content of about 4% [18]. The decrease of other properties for the first 1% of voids is reported as high as 30% (flexural strength), 9% (torsional shear), 8% (impact strength) and 3% (tensile properties).

These defects just described are the mostly recurrent ones; but, there are also some defects that are peculiar to specific manufacturing processes.

In particular, composites fabricated with pre-impregnated layers may entail some specific defects due to the improper storage of pre-pregs, like:

- out-of-date resin because of exposure to ambient temperature (higher than that required for correct storage)
- wrinkled surface because of an uneven position, which may result in resin-rich regions within the laminate
- accumulation of debris, resulting in slag inclusions
- broken, or damaged, fibre tows resulting in reduced strength of the laminate

During the processing in an autoclave, some defects can occur in the laminate, mainly related to imperfectly cured resin regions due to incorrect pressure/temperature reached values. Besides, lack of bonding between layers, due to nonuniform pressure on the whole surface, may also occur.

Typical drawbacks of RTM are the formation of porosity due to the volatilization of dissolved gases in the resin, the mechanical entrapment of gas bubbles or a possible evaporation of mould-release agents. The void formation can be influenced by many factors, such as resin properties, moulding temperature, injection pressure and external pressure during curing.

Of course, whatever the type of defect, it may result in slight, or severe, variations of the material properties with, in turn, a reduction of the predicted life. In fact, these variations may act as sites for the initiation of fatigue damage or may facilitate the growth of a fatigue crack during cyclic loading. A comprehensive assessment of the quality of a composite material prior to putting it into service is therefore as important as the monitoring of the levels of damage accumulated in a composite structure during service [19]. This justifies the increasing attention toward development of:

- effective nondestructive evaluation methods able to discover defects at an incipient stage
- aerospace design procedures aimed at achieving zero-growth thresholds for any type of defect

Theoretically many defects may be avoided and the overall quality of a component may be increased with the implementation of particular procedures and the use of specific instrumentation. As an example, the use of computer-controlled tape-laying machines may assure construction of a pre-preg stack for autoclaving to very high standards of quality and repeatability. Similarly, errors of control in pressing can be avoided to some extent by the use of automated autoclaves with pressure—temperature cycles carefully programmed from detailed chemical knowledge of the gelation and viscosity characteristics of the resin in use [19]. Of course, this entails extra costs that remain an industrial decision.

A challenging question may arise concerning the real consequences of the presence of a defect. As stated by Harris in his comprehensive book [19], a few isolated spherical pores, a micron or so in diameter, can have no effect on any physical property, and may not therefore affect the tensile or flexural mechanical performance of a composite. But a distribution of innocuous-looking pores can markedly reduce the interlaminar shear strength of a material and, by providing sites for accumulation of moisture, may also decay the electrical or dielectric performance of the material. A minor delamination between plies in a complex laminate may have no effect on the tensile strength of the material, but such defects are frequently injurious to the compression performance of laminates and, as indicated previously, can rapidly grow to eventually damage the composite under cyclic-loading conditions. Therefore, to avoid unpredictable failure, it is important to use the most effective nondestructive defects detection techniques and then establish the critical size with a case-by-case approach owing to the specific type of composite and the specific service conditions for the given application.

1.8.2 In-Service Failures

The in-service failure of composite aircraft structures is primarily caused by impact strikes [20]. The weakness of composites to impacts is well known, especially to low-energy impacts that cause subsurface damage in the form of delamination, cracks as well as fibre breakage. These impacts show little visual evidence on the external surface as to the extent of the delaminated area other than a small dent [21]. In particular, delamination in CFRP laminates reduces the strength, particularly the compression strength, of the structure; if undetected and unrepaired, it may result in reduction of ultimate load capability or even inability to withstand design limit load [22]. Conversely, for safety issues, when impact damage occurs to an aircraft structure, the structure must be able to 'withstand reasonable loads without failure or excessive structural deformation' for the 'operational life of the aircraft' or 'until the damage is detected' [23].

Of course, for design purposes, it is of great importance to have knowledge of the failure mechanisms. In the Griffith model [24], the fracture of homogeneous materials is based on the stress-intensity factor, K, and the associated strain-energy release rate, G, which is mainly related to the material fracture toughness (material resistance to crack propagation) [19]. In real fibre composites, the microstructural inhomogeneity and anisotropy cause the failure process to be very complex with also a combination of microfailure events, which can often give rise to high levels of fracture energy. In fact, unlike homogeneous materials like metals, composites have crack-stopper ability inherent in both types of interfaces at a microscopic level (between fibres and matrix) and at a macroscopic level (between separate laminas). This gives rise to complex fracture mechanisms involving breaking of fibres and matrix, delamination between fibres and matrix and a combination of crack deviations along interfaces (at both micro- and macroscopic levels). In practice, the fracturing of a composite is driven by three factors: the matrix, the reinforcement and the interface; it is important to consider the types of matrix and fibres, their mutual volumes and the type of bonds in between them, meaning the curing process. Then, the toughness of a composite is derived from many sources, and the relative magnitudes of the separate contributions depend not only on the characteristics of the components but also on the manner in which they interact. There is thus no simple recipe for predicting the toughness of all composites [19]. It follows that the procedures based on classical fracture mechanics cannot be simply applied for design purposes of composites. When fibres are incorporated into a matrix of any kind, the separate phases may be able to contribute their individual levels of toughness to the composite in an additive fashion, or they may not, depending on the kind of interaction that occurs and on the level of constraint that is set up as a result of the differences in their properties [19]; so, it is rather difficult, or even impossible, to predict the way a composite will fail.

Since the introduction of composites in aircraft construction, the main interest of the scientific community was to establish the delamination threshold load (DTL) under impact [25]. However, notwithstanding the huge amount of available data coming from both numerical simulation and experimental testing, a methodology to unambiguously establish the DTL has still not been completely achieved. This is because the DTL depends on many factors, in primis, the effective material mechanical characteristics as well as the geometry of the target [26,27] and of the impactor [28].

The high variability of mechanical properties of composites, as a main consequence of porosity that is induced by manufacturing processes and that is practically unavoidable [29], is well known; in fact, the voids content can be reduced but not completely eliminated. Perhaps this is the reason why composite materials display a large variety of damaging ways under impact. In fact, as reported by Abrate [30], some impacts produce deformations in a small zone surrounding the point of impact while others involve deformations of the entire structure. In some cases, a major portion of the impact energy is transferred to the plate, and in other cases, most of the energy is returned to the impactor. For some problems, the indentation absorbs a significant portion of the impact energy so that it must be adequately modelled in the analysis; in other cases, the effects of indentation are practically negligible.

It must also be considered that mainly in the aircraft industry there is the habit to quantify impact threats in terms of impact energy, but this, as observed by Olsson et al. [31], is not at all correct since a small mass and a large mass impactor entail, at the same impact energy, completely different responses. A solution to this drawback was recently devised by Meola and Carlomagno [32] who proposed a relationship that links the damaged area to the impact energy and the impactor dimension. This result was achieved by the authors while using infrared thermography to investigate the response of composites to impact events [32,33]. In particular, they proved that, with an infrared imaging device, it is possible to visualize the thermal effects that develop under low-energy impact, and which may supply information about initiation and propagation of the impact damage. By considering that during an impact event kinetic energy passes from the impactor nose to the target, and that such energy is in part dissipated as heat, the detection of thermal signatures developing under impact is important for the comprehension of failure modes. In fact, any form of damage (delamination and/or fibre breakage) is generally accompanied by heat dissipation, which manifests itself through the appearance of hot spots/areas over the material surface. In this regard, the use of infrared thermography has to be considered as beneficial and unique; this topic will be addressed in more detail in a following section.

1.9 TOWARD NOVEL TECHNOLOGIES

As already mentioned, the development of new materials of enhanced properties such as resistance to high temperature, high stiffness coupled with high lightness, have always been a primary objective in the aeronautical field. In particular, the exploitation of novel materials, in both military and civil aviation as well as space aeronautics, have allowed achieving results that were impossible to obtain with traditional materials [34]. The development of new materials is constantly progressing, and now, in the era of composites, it is rather easy to create a new material by simply adding a new ingredient to the basic recipe or by changing the curing parameters. In particular, presently (in which words like *smaller, younger, mini* and *thinner* are an imperative) the word *nano* is ever more dominating the composite materials world.

Indeed, nanotechnology allows control of matter and processes at the nanoscale (1–100 nm) [35,36] with the possibility to create materials of desired characteristics. The use of nanomaterials as filler in polymer resins has the advantage of a high aspect ratio with the main effect of an increase in the reinforcement surface area. This offers the possibility to obtain the same performance properties achieved with traditionally filled resins but with a smaller filler volume fraction. In addition, nanofilled resins often exhibit other novel beneficial characteristics on the macroscale properties of the composite such as improved thermal and electrical conductivity, reduced flammability, higher stiffness, strength and resistance to wear and damage. However, the manufacturing of nanocomposites has raised the challenge of the thorough dispersion of nanoscale fillers throughout the resin.

A useful application of nanotechnology has been devised in the enhancement of fibre-reinforced polymer composites. This may include three approaches [37]:

1. modification of the matrix
2. modification of the reinforcement

3. incorporation of a macroscopic arrangement of a nanometre-dimensional material

Of course, the new material may be created by pursuing one, or more, of the aforementioned three approaches. In particular, a novel technology that is still under development [38] consists of the production of reinforcements (glass or carbon) infused with carbon nanostructures to be used as pre-pregs for the fabrication of composites. This technology is known as carbon-enhanced reinforcement and it is expected to deliver improved composite in-plane shear strength and greater interlaminar shear strength.

DISCLOSURE

Some figures, which refer to well known facts, were arranged traying inspiration from websites.

SUMMARY TO CHAPTER 1 AND INTRODUCTION TO CHAPTER 2

The aim of this book is to make the readers aware of the usefulness of infrared thermography with composite materials, including design, development, evaluation of soundness and maintenance.

Within this context, this first chapter has to be regarded as an introduction to composites. The attention has been mainly focused on the composites used for aircraft construction. Therefore, some basic information is provided involving the main types of composites with varying of the type of matrix and of reinforcement, as well as some hints on the fabrication processes. A section is dedicated to descriptions of the main types of defects that may arise in composites, either during manufacturing or in service. From such descriptions it is possible to understand the main weaknesses of composites. The chapter ends with a look at the future, which appears to be dominated by nanotechnology.

Chapter 2 is entirely focused on the assessment of the material soundness in a nondestructive way. In particular, many different nondestructive techniques are described and analyzed, owing to their application to composites. This is performed with the primary intention to make the reader aware of the availability of techniques and methodologies and provide the elements for making an informed choice. However, the matters discussed in Chapter 2 may also help as the basis for determining, through a general comparison, whether infrared thermography may be used in place of, or integrated with, the actual most commonly used techniques. This, of course, is in order to take advantage of the absence of contact with the surface and the possibility of remote control, which is inherent in any test carried out with infrared thermography.

REFERENCES

[1] P.L. Jakab, Wood to metal: the structural origins of the modern airplane, J. Aircraft 36 (6) (1999) 914–918.
[2] D. Hull, T.W. Clyne, An Introduction to Composite Materials, Cambridge University Press, 1996.
[3] C. Soutis, Fiber reinforced composite in aircraft costruction, Progr. Aerosp. Sci. 41 (2005) 143–151.
[4] D.J. Landini, R.L.K. Matsumoto, D.M. Domanski, Processing of Sic/Sic Ceramic Matrix Composites by Use of Colloidal Carbon Black, US 0035024 A1, 2006.
[5] K.U. Kainer, Metal Matrix Composites. Custom-made Materials for Automotive and Aerospace Engineering, Wiley-VCH, 2006.
[6] D.B. Miracle, B. Maruyama, Metal matrix composites for space systems: current uses and future opportunities, in: M. Strophy (Ed.), Proceedings of National Space and Missile Materials Symposium San Diego (CA), February 28–March 2, Ankon Corp, Dayton, OH, 2000.
[7] Dr. M.A. Masuelli, Introduction of fibre-reinforced polymers – polymers and composites: concepts, properties and processes, in: M. Masuelli (Ed.), Fiber Reinforced Polymers - the Technology Applied for Concrete Repair, InTech, 2013, ISBN 978-953-51-0938-9, http://dx.doi.org/10.5772/54629.
[8] C.M. Pastore, D.W. Whyte, H. Soebruto, F.K. Ko, Design and analysis of multiaxial warp knit fabrics for composites, J. Ind. Fabrics 5 (14) (1986) 4–14.
[9] B.N. Cox, G. Flanagan, NASA Contractor Report 4750. Handbook of Analytical Methods for Textile Composites, 1997.
[10] F.T. Wallenberger, J.C. Watson, H. Li, in: Glass Fibers ASM Handbook, Composites, vol. 21, 2001, pp. 27–34.
[11] A.C. Long (Ed.), Design and Manufacture of Textile Composites, Woodhead Publishing and CRC Press LLC, 2005, ISBN 9781855737440.
[12] M.E. Greenwood, J.V. Gauchel, J.J. Beckman, A.G. Hankin, Process for Filament Winding Composite Workpieces, US 6179945, 2001.
[13] T. Sekido, K. Kitaoka, H. Odani, S. Nishiyama, M. Shimizu, Method of RTM Molding, US 0125155 A1, 2006.
[14] L.E. Letterman, Resin Film Infusion Process and Apparatus, US 4622091, 1986.
[15] A. Vlot, J.W. Gunnink (Eds.), Fibre Metal Laminates: An Introduction, Kluwer Academic Publishers, The Netherlands, 2001. ISBN:I-4020-0038-3.
[16] J. Summerscales, Manufacturing defects in fibre reinforced plastics composites, Insight 36 (12) (1994) 936–942.
[17] H.T. Yoshida, T. Ogasa, R. Hayashi, Statistical approach to the relationship between ILSS and void content of CFRP, Comput. Sci. Tech. 25 (1) (1986) 3–18.
[18] N.C.W. Judd, W.W. Wright, Voids and their effects on the mechanical properties of composites - an appraisal, SAMPE J. 14 (1) (1978) 10–14. January/February.
[19] B. Harris, Engineering Composite Materials, The Institute of Materials (IOM), London, 1999. ISBN:1861250320, 9781861250322.

[20] Civil Aviation Authority Safety Regulation Group, Reliability of Damage Detection in Advanced Composite Aircraft Structures Paper 2013/03, 2013.
[21] M.S. Sohn, X.Z. Hua, J.K. Kimb, L. Walker, Impact damage characterisation of carbon fibre/epoxy composites with multi-layer reinforcement, Composites: Part B 31 (2000) 681−691.
[22] Transportation Safety Board of Canada, Assessment of the Response from Transport Canada to Aviation Safety Recommendation A06-05 Inspection Program of Rudder Assembly, 2007.
[23] European Aviation Safety Agency, Certification Specifications for Large Aeroplanes CS-25, Amendment 5, 2008.
[24] A.A. Griffith, The phenomena of rupture and flow in solids, Philos. Trans. R. Soc. Lond. Ser. A 221 (1920) 163−198.
[25] D.J. Elder, R.S. Thomson, M.Q. Nguyen, M.L. Scott, Review of delamination predictive methods for low speed impact of composite laminates, Compos. Struct. 66 (2004) 677−683.
[26] G.A. Schoeppner, S. Abrate, Delamination threshold loads for low velocity impact on composite laminates, Composites: Part A 31 (2000) 903−915.
[27] W.J. Cantwell, Geometrical effects in the low velocity impact response of GFRP, Compos. Sci. Technol. 67 (2007) 1900−1908.
[28] T. Mitrevski, I.H. Marshall, R. Thomson, R. Jones, B. Whittingham, The effect of impactor shape on the impact response of composite laminates, Compos. Struct. 67 (2005) 139−148.
[29] L. Liu, B.M. Zhang, D.F. Wang, Z.J. Wu, Effects of cure cycles on void content and mechanical properties of composite laminates, Compos. Struct. 73 (2006) 303−309.
[30] S. Abrate, Modeling of impacts on composite structures, Compos. Struct. 51 (2001) 129−138.
[31] R. Olsson, M.V. Donadon, B.G. Falzon, Delamination threshold load for dynamic impact on plates, Int. J. Solids Struct. 43 (2006) 3124−3141.
[32] C. Meola, G.M. Carlomagno, Impact damage in GFRP: new insights with infrared thermography, Composites: Part A 41 (2010) 1839−1847.
[33] C. Meola, G.M. Carlomagno, F. Ricci, V. Lopresto, G. Caprino, Investigation of impact damage in composites with infrared thermography, in: P. Mazal (Ed.), Proceedings of 6th NDT in Progress, Prague (Czech Republic), 2011, pp. 175−182.
[34] A.A. Baker, S. Dutton, D. Kelly, Composite Materials for Aircraft Structures, second ed., AIAA Education Series, 2004.
[35] K.J. Klabunde, J.V. Stark, O. Koper, C. Mohs, A. Khaleel, G. Glavee, D. Zhang, C.M. Sorensen, G.C. Hadjipanayis, in: C. Hadjipanoyis, R.W. Siegel (Eds.), Chemical Synthesis of Nanophase Materials, Nanophase Materials, Kluwer Academic Publisher, The Netherlands, 1994, pp. 1−19.
[36] D.R. Paul, L.M. Robeson, Polymer nanotechnology: nanocomposites, Polymer 49 (2008) 3187−3204.
[37] M. Bauera, O. Kahleb, S. Landeckc, C. Uhligd, R. Wurzel, High performance composites using nanotechnology, Adv. Mater. Res. 32 (2008) 149−152.
[38] http://www.compositesworld.com/articles/nanotechnology-into-the-realm-of-real (accessed on February 2014).

CHAPTER 2

Nondestructive Evaluation

2.1 APPROACHING THE NONDESTRUCTIVE WORLD

To be able to see through matter, to ascertain if a final product is free of hidden defects without destroying it, has certainly been the ultimate desire of every manufacturer! Unfortunately, our eyes work in a limited band of the electromagnetic spectrum (about 0.39–0.75 μm), which is called the *visible band*, and only in the presence of light; in fact, we cannot see in the darkness. Indeed, what we really see is the light reflected by an object, which forms an image on the retina of our eyes. And then, through the human vision, it is only possible to perceive the presence of defects of a certain size, located on the surface of the object, unless such an object is transparent to the visible electromagnetic band, eg, like glass. The vast majority of useable substances are not transparent in the visible band, thus hindering our vision of hidden defects. However, thanks to technological progress, we have only to apply some physical principles that make it possible to travel inside matter and see what is invisible to the naked eye.

Before embarking on discussion of the nondestructive evaluation (NDE) of composite materials, which is the primary focus of this book, it is essential to first establish the correct semantics to be used. In fact, terms like *nondestructive inspection* (NDI), *nondestructive testing* (NDT) and NDE are encountered in the literature and often they are used with almost the same significance. While at first they may appear to be synonyms, with deeper analysis and from a more rigorous point of view fundamental differences may be found between the three terms *inspection*, *testing* and *evaluation*. Then, the question arises of what is the correct term to use? To shed some light on this matter, we will analyze each term separately.

In a general sense, the term *inspection* indicates the formal act of carefully and critically viewing something or somebody. The subject of the inspection may be different involving either an activity or a product, with the prerequisite being almost the same, which is to be in line with specific standards. It is mainly a job for the mind that primarily involves the sense of sight (ie, the eyes). To give some examples, the Food Safety Inspection Service is charged with ensuring that foodstuffs are safe, through their entire

Infrared Thermography in the Evaluation of Aerospace Composite Materials
ISBN 978-1-78242-171-9
http://dx.doi.org/10.1016/B978-1-78242-171-9.00002-4

life cycle, which involves raising of crops and livestock, food processing, warehousing, transportation and distribution to the final consumer. In particular, there are national and international agencies (dedicated to the safeguarding of foods, animals and plants) to enhance the health and well-being of the people, environment and economy. The inspection of vehicles is concerned with their performance, conformity with laws regarding safety of passengers, release of pollutants into the atmosphere etc. Again, it is inspection of any assets in the public administration, such as documents or facilities, records to assure compliance with regulations. A medical inspection is a thorough and unhurried examination of a patient to check his or her health conditions. Inspection of a property is performed to check the conditions of houses, or land, and the existence of any restrictions before purchasing them. In some engineering endeavours, inspection may involve measurements, tests and gauges applied to certain characteristics in regard to an object, or activity, but the results are usually compared to specified requirements and standards for determining whether the item, or activity, is in line with these targets. In Europe, inspection may be assessed by accreditation bodies according to the ISO/IEC 17020 standard, which defines inspection as: 'examination of a product, process, service, or installation or their design and determination of its conformity with specific requirements or, on the basis of professional judgment, with general requirements' [1]. The inspection may be accompanied by a possible legal action as a consequence.

The word *testing* indicates mainly an action — do something to ascertain if someone, or something, has certain features. It involves work of arms and mind. This may apply to different sectors with different means. A set of questions, problems, or the like, is used for evaluating the abilities, aptitudes, skills or performance of human beings. Testing in chemistry is mainly the process of identifying the constituents of a certain substance. Similarly in agriculture, a soil test is made to determine the acidity or pH level, as well the composition in terms of nutrients and contaminated content. A machine is tested to assess its performance, or even better to see if it complies with the design purposes. Often, machines are used to test materials and other machines. Software testing is carried out to verify that a computer program meets the requirements that guided its design and development, works as expected and satisfies the needs of stakeholders. In any case, whatever the testing type, it is a procedure for critical evaluation, a means of determining the presence, quality or truth of something in something else.

The term *evaluation* takes on the meaning of expressing an opinion. The subject under evaluation may be either an individual (human or animal), an organization, an activity, a product or simply an idea. The output is a systematic determination of the subject's merit, worth and significance. It is mainly a work of mind and it is based on the competence and expertise of evaluators; of course, professional ethics are also of great concern. Often evaluation is directly connected to testing, or better, follows the testing phase since the output of a test may either be a matter of evaluation, or helps in decision-making. Mostly, evaluation is inherently a theoretically informed approach to interpret and give meaning to the results obtained from tests.

As an important point, it can be assessed that, also when coupled with the word nondestructive, the three terms preserve their principal meaning. And so NDI mainly entails vision in a general sense, which may lead either to discard a component, if macroscopic nonconformities are observed, or to proceed with the execution of specific tests (nondestructive tests) in order to gain more information. Nondestructive testing involves the physical action performed on the subject to verify/understand its characteristics, behaviour, value, etc.; the output of such tests deserves particular attention since test results shall be critically analyzed and evaluated. And finally, NDE represents the action of making, on the basis of testing results, the final judgement. Thus, going back to a composite material, to express an opinion on its condition it is important to go through testing and evaluation. In this book, the term NDT is used to refer only to tests, while the term NDE is used to involve both testing and evaluation.

But, what to discover? What to search for?

At the end of any nondestructive event, what is important is the final decision! Or better, the response provided to enquiries like these:

- Is the part free of defects?
- Needs repair?
- Must be anyhow rejected?

Of course, a straight answer on the spot cannot be delivered, but several factors must be taken into consideration. An important role in the decision-making is played by the intended use of the part and the acceptance criteria; the latter may be different depending on the specific sector (industry) and destination. Generally, whatever the industry, acceptance criteria are based on the presence, nature and size of defects. This would mean that in some cases no defects, even if very small, are acceptable, and the part must be discarded if any impurity is perceived. In other cases, the part passes the control if it contains defects at less than a fixed level.

2.2 DEFECT TYPES

In general, the word *defect* is used to mean something wrong or undesired in an artefact or a structure. In fact, it comes from the Latin word *defectus*, which means deficiency, weakness or lack of something for completeness. A material may have several types of defects, which can be largely grouped into two main categories, namely micro- and macrodefects.

Microdefects are all the defects that affect the material crystalline structure and are also called *crystalline defects*. Crystalline defects may also be regarded as false defects since they give the material specific properties and, then, they can also be exploited for technological purposes. More specifically, defects in the crystal structure may include either spoiled atoms or irregularities in the atoms' alignment, and may be classified as point defects or linear defects.

- Point defects are local imperfections in the crystal, like:
  - The missing of an atom, which means the presence of a vacancy or a hole (Fig. 2.1(b)); a concentration of holes in metals can substantially vary their electric and thermal conductivities.
  - The presence of an interstitial impurity (Fig. 2.1(c)). Interstitial impurity is an atom of a different type than the atoms of the bulk matrix. An example is the carbon atoms that are added to iron to make steel.
- Linear defects are misaligned planes of atoms that are present in the crystal structure and are called *dislocations*.

Macrodefects are real, undesired defects. Generally, they represent a discontinuity in the material, which may form during manufacturing and/or in service. Incorrect manufacturing processes may cause formation of defects such as voids, porosity, cracks, foreign inclusions etc. In service,

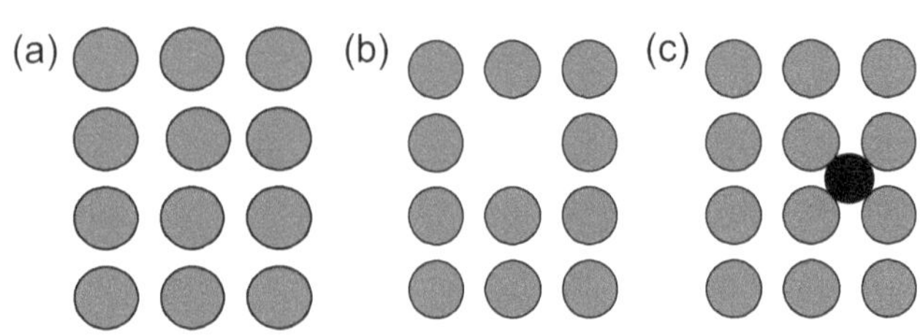

Figure 2.1 Crystalline defects. (a) Normal, (b) missing atom and (c) interstitial defect.

material degradation occurs as mainly due to fatigue cycles, impact damage and adverse environmental conditions that cause corrosion and overall decay.

Defects may be, in general, classified as:

- open defects such as cracks and voids
- gas-filled volumes like bubbles
- solid slag inclusions (backing film, dust, foreign material fragments)
- lack, or excess, of adhesive in bonded joints and in adhesive-based structures

Examples of some macroscopic defects are sketched in Fig. 2.2. The first scheme (Fig. 2.2(a)) shows a general flaw that may be either a solid slag inclusion or a gas-filled volume; real defects are in general of complex geometry with jagged edges. The second scheme (Fig. 2.2(b)) represents a discontinuity in the adhesive film between two surfaces of a junction. This type of discontinuity is generally very thin and small since it is located over either narrow or large surfaces but in a spotty fashion.

Of course, a complete classification of defects has to be performed owing to specific types of materials and structures and to the specific field in which the given material is employed (transport industry, architecture, environment etc.). This is because some defects have no effects on the material's functioning and life, while others may enlarge in service and can seriously affect the material performance and the component's working life. The latter are defects that must be discovered in time, before catastrophic consequences! Of course, most important is the use that the material is intended for, or better, whether or not the material is used for the construction of structural parts.

Once one understands that it is important to discover hidden defects, the question that arises is, how to discover all these buried defects without destroying the material? In the following section, we try to provide a response.

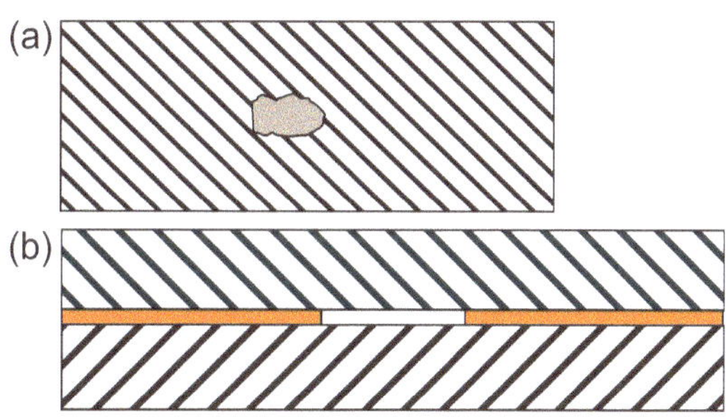

Figure 2.2 Examples of macroscopic defects. (a) Flaw buried in the material and (b) lack of adhesive between parts of a joint.

2.3 TECHNIQUES AND PROCEDURES: SOME GENERAL HINTS

Once a part has been manufactured, it is important to ascertain its conformity with specific standards and the absence of any buried defect. The oldest and cheapest way a man used to control the conformity of an artefact was by his senses, which means visual inspection, percussion and auscultation. Of course, visual inspection is perhaps the method every manufacturer uses automatically, without thinking of it as a particular method, to verify the artefact's correctness. In other words, a look at the finished object permits checking shape, dimensions and surface conditions in order to reject parts that are wholly out of order. Percussion is mainly used in architecture to individuate areas of detached plaster; the bond quality is estimated from the acoustical response to a knock. Of course, results are almost subjective because they depend mostly on the operator's skill and experience. Conversely, the results of a quality control must be quantitative and of overall validity (ie, objective).

Today many different techniques are available to control quality, and yet, as technology evolves and novel materials are created and new types of defects need detection, more sophisticated devices are constantly developed too, which make possible the development of new techniques as well the enhancement of the existing ones. The existence of different techniques means that many different physical parameters can be exploited for NDE of materials. In fact, there are techniques that are based on thermal-, acoustic- and electric-signal type, which may be used for detection of many types of defects, whether they are shallow or deep. Roughly speaking, NDE can be performed by launching an energy input into the material and analyzing its response. More specifically, a certain signal (which might be electric current, heat flux, ultrasound waves, X-rays or something else) is delivered to the material surface, then either the material reaction to such a stimulus or the modifications experienced by the signal itself through the material are observed and analyzed.

In general, when a given signal touches a material surface, it may be absorbed, reflected or transmitted according to the material reflectivity, absorptivity and transmissivity for such a signal. The propagation of the signal inside the material depends on its inherent mechanical, thermal, chemical, acoustical, electrical and optical properties. Of course, such properties modify in the presence of inhomogeneities and overall material alteration. As an important consequence, it is possible to evaluate the variation of a material property through the variation of another property. All one needs to do is to choose the most appropriate technique, taking into account that its effectiveness depends on two main factors:

- The properties of the host material (thermal, optical, electrical etc.) to let the signal travel through.
- The perturbation that a hidden discontinuity causes to the propagation of the specific test energy signal.

In other words, the readable quantity, related to the induced perturbation, must be large enough to not be confused with the background noise. Owing to its inherent characteristics, a material can be evaluated by using the most effective signal-based technique or a combination of techniques. On the whole, an inspection involves two main aspects: the material degradation in terms of loss of original characteristics (variation of density, hardness, elasticity, thermal or electrical conductivity etc.) and the presence of defects both superficial (scratches, loss of optical properties etc.) and internal (voids, delaminations, cracks etc.). Basically, no technique is able to detect every type of defect in every type of material, and thus the most effective technique for the specific application must be selected. Sometimes a synergic integration of several techniques is compulsory.

2.3.1 Classification of Techniques

All the available techniques can be classified depending on the way a test is carried out, or better, on the relative position of the energy input actuator and/or the sensor with respect to the surface of the component under test. Then, it is possible to arrange all the techniques into the two groups of *contact* and *noncontact*:

- *Contact*: all the techniques that involve either direct contact with the surface of the inspected component or an indirect contact when a coupling medium is needed.
- *Noncontact*: no direct contact, nor coupling medium, is required; the testing is carried out by stimulating the object from a remote position.

Another classification may be attempted that is based on the type of output signal involved, which may be thermal energy, sound, X-rays, electric and magnetic fields etc. The main signal classes are the following:

- *Thermal*: heat is delivered to the surface of the component or it is generated inside the material. Then, the analysis is performed as a result of heat propagation and/or local variations of temperature.

- *Sound*: sound waves, often ultrasound waves, are exploited for the material evaluation owing to the changes that the signal undergoes and its crossing speed.
- *Electric*: this class includes all methods that use either a direct electric current injected into the material under test or any effects related to the electric/magnetic field. In our discussion, the magnetic-based techniques are forced to fall into the electric class. Indeed, electric and magnetic fields are different and may exist independently, but they are also interrelated as the magnetic field is produced by moving electric charges.
- *Radiation*: short-wavelength (usually, less than 10 nm) electromagnetic radiation, like X-rays, gamma rays or neutrons, are used to penetrate an object of a given material. The radiation emerging from the inspected object's opposite site is analyzed to search for any variation of intensity (amount) that may be related to the material properties.

Of course, this is a quite general way to group several different nondestructive techniques mainly using the output (reading) signal. In fact, heat may also be generated by an electric current through the Joule effect, or by sound waves through the friction effect, involving different types of output signals. In any case, the classification is based on the output signal that is directly under analysis, without any distinction of what caused it.

A final classification may be done considering the safety issues of the workplace and emissions that may be detrimental to the environment:

- *Harmful*: all the techniques that include any potential health risk, whether they are mobile mechanical parts, chemicals, radioactive emissions, electromagnetic radiation etc. X-rays represent an example of a harmful technique. Of course, dangerousness also concerns emission of substances harmful to the environment (ie, release of pollutants into the surroundings).
- *Harmless*: to be completely harmless, a technique must not emit harmful radiation, cannot require exposure to hazardous substances and must not include any harmful part in the test set-up.

Of course, the optimum is reached by choosing the best from each of the three classification schemes described herein. As depicted in Fig. 2.3, a good NDT technique should include different characteristics.

- *Defects detection*: this entails the possibility to discover anomalies at an early stage before they manifest negative effects on the part performance. This, of course, entails high resolution, high sensitivity, or better, often high output variations of the input signal.

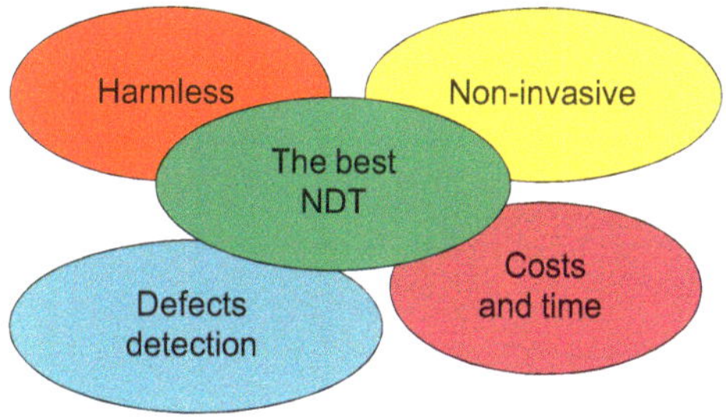

Figure 2.3 A good nondestructive testing choice.

- *Noninvasive*: this is principally related to the part under evaluation, meaning that care is paid to safeguard the inspected part and to avoid any alteration that may affect its performance.
- *Harmless*: this aspect is mostly devoted to safeguarding the human beings. Of course, this involves the safety of the operator (ie, workers directly involved with the inspection) as well as environmental protection, without release of pollutants, which, in turn, have negative effects on other humans' health.
- *Costs and time*: these are important parameters in the manufacturing industry that can pose severe limitations on the use of a particular technique.

However, being aware of the type of material to evaluate and the type of defect to discover, the choice of the best technique should be made through a fair compromise between the four points just discussed.

2.4 ABOUT APPLICATION OF NONDESTRUCTIVE TESTING TECHNIQUES

In general, NDT activities are also, somehow, operator dependent; therefore, successful and consistent application of NDT techniques depends heavily on personnel skill. Those in authority are asked to put great reliance on training, experience, judgment and integrity of the involved personnel [2]. Consequently, today globalization asks for certification harmonization and recognition of NDT personnel, all over the globe. The first step in doing this is the identification of common standards.

Standards always represent an effort by some organized groups of people to improve the consistency of manufacturing materials and processes, as well as the resulting products. A standard specification is the result of an agreement among the involved parties and usually entails acceptance of its use by them. It includes a detailed description of how to produce something or how to perform a particular task. Anytime a product is marked as meeting a specification, or a contract requires the use of a specification, the product or service must meet the requirements of that document. Standard specifications are of great value to industry and offer many advantages. The most important one is perhaps the possibility for the manufacturer to introduce production standards that, in turn, allow for mass production with reduction of costs. In addition, the use of standards offers some benefits also to the consumer in terms of expected product performance. Finally, the use of standards allows the comparison of results coming from tests in different laboratories worldwide.

Conversely, a disadvantage of standard specifications may be the likelihood of freezing practices that are based on poor knowledge, slowing the development of better practices. However, it must be remembered that standards do not imply a degree of perpetuity (like dimensional or other reference standards); on the contrary, technical advances in a given field usually call for periodic revision of the enforced requirements.

2.4.1 Some Historical Dates in Standards Development

As a first milestone, the American Society for Testing and Materials (ASTM) was founded in 1898; it now provides an international forum for the development and publication of standards for materials, products, systems and services used in construction, manufacturing and transportation. On a wider basis, in 1947 the International Organization for Standardization (ISO) was formed as a nongovernmental federation of standardization bodies from over 60 countries with headquarters in Geneva, Switzerland. As of 2013, ISO works in 164 countries.

However, the boost in the development of remarkable and very accurate standards has come principally from the world of aerospace in which quality and safety are primary issues. In the United States, the Aerospace Industries Association (AIA) represents the nation's major manufacturers of commercial, military and business aircraft, helicopters, aircraft engines, missiles, spacecraft, materials, and related components and equipment. The AIA, originally known as the Aeronautical Chamber of Commerce (ACCA), has been an aerospace industry trade association since 1919. It is responsible for the document named NAS 410 - Certification and qualification of NDT personnel, formerly known as MIL-STD-410E (Military Standard for NDT personnel qualification and certification), which establishes the minimum requirements for the qualification and certification for personnel involved in NDT. These requirements include training, experience and examination.

In Europe, qualification and certification of personnel involved in NDT for the aerospace manufacturing sector is performed in accordance with EN4179 [3], adopted by the European Aviation Safety Agency (EASA). EASA is established by the national Civil Aviation Authorities. More specifically, in many countries there is a National Aerospace NDT Board (NANDTB), representing the nation aerospace industry, which provides guidance to NDT qualification and examination services in accordance with EN 4179.

In the light of international harmonization of standards, it is worth mentioning the agreement reached by EASA and the United States Federal Aviation Administration. This was addressed through the revision of NAS 410 in February 2003 to set equivalence in technical content with the EN4179 (2000). In addition, the new EN4179 [4] dictated technical equivalence with the NAS 410 (February 2003). Later, in 2008 the EN 4179 was revised to achieve technical harmonization with the revised AIA-NAS-410 and so on. It is a common practice to use the term NAS410/EN4179, or EN4179/NAS410, to refer to guidelines for qualification and certification/approval of NDT personnel.

2.4.2 Qualification and Certification of Personnel for Nondestructive Testing

Personnel involved in application of industrial NDT methods and interpretation of results should be certified, and in some industrial sectors (like aerospace), certification is enforced by law or by the applied codes and standards [5].

It is necessary to distinguish between qualification and certification.

Qualification: demonstration of physical attributes, knowledge, skill, training and experience required to properly perform NDT tasks.

Certification: Procedure, used by the certification body to confirm that the qualification requirements for a method, level and sector have been fulfilled, leading to the issuing of a certificate.

Nondestructive testing training is provided for people working in many industries. It is generally necessary that the candidate successfully completes a theoretical and practical training program, as well as performs several hundred hours of practical application of the particular method for which he wishes to be qualified. At this point, he may pass a certification examination. While online training has become very popular, many certifying bodies still require additional practical training.

Basically there are two approaches in personnel certification [6]:

1. *Employer-based certification*: the employer compiles their own written practice. In US industrial sectors, the written practices are usually based on recommended practice SNT-TC-1A of the American Society for Nondestructive Testing (ASNT) [7].
2. *Personal central certification*: within this approach NDT operators obtain certification from a central certification authority, which is recognized by most employers, third parties and/or government authorities. Industrial standards for central certification schemes include ISO 9712 [8] and ANSI/ASNT CP-106 [9]. Another central certification scheme, very similar to ISO 9712, is the EN 473 [10], which was withdrawn when the European Committee for Standardization replaced it with EN ISO 9712 in 2012.

In the United States, employer-based schemes are the norm, however, central certification schemes exist as well. In the European Union, Level III central certification (see later discussion) is more widely used, where certifications are issued by accredited bodies (independent organizations conforming to ISO 17024 and accredited by a national accreditation authority).

The aerospace sector worldwide sticks to employer-based schemes, as described in the previous paragraph, and based mostly on AIA-NAS-410 [11] in the United States and on the equivalent and very similar standard EN 4179 [12] in the European Union. However, the EN 4179 (2009) includes an option for central qualification and certification by a NANDTB.

Unlike the certification approach used, three levels of qualification and/or certification are generally used, namely: Level 1, Level 2 and Level 3. The roles and responsibilities of personnel for each level are generally as follows:

Level 1: technicians qualified to perform only specific calibrations and tests under close supervision and direction of higher-level personnel. Normally, they work following specific work instructions for testing procedures and rejection criteria and they can only report test results.

Level 2: engineers, or experienced technicians, who are able to set up and calibrate testing equipment, conduct the inspection according to codes and standards (instead of following work instructions) and compile work instructions for Level 1 technicians; they can supervise and train Level 1 technicians as well. They are also authorized to report, interpret, evaluate and document testing results. In addition to testing methods, they must be familiar with applicable codes and standards and have some knowledge of the manufacturing and service of tested products.

Level 3: usually specialized engineers or very experienced technicians. They can establish NDT techniques and procedures and interpret codes and standards. They may be responsible for NDT laboratories and have a central role in personnel certification. They are expected to have wider knowledge that covers materials, fabrication and product technology.

On the whole, there are small differences between different codes and standards, regarding personnel responsibility.

In 1966 the ASNT published the first edition (ASNT SNT-TC-1A) of the Recommended Practice to provide guidelines for the personnel involved in nondestructive tests. This document underwent several editions until 2006 to include novel techniques. In fact, the first edition involved five methods, including ultrasonic testing, magnetic testing, liquid penetrant testing, eddy current testing and radiographic testing. By 2006 the techniques increased to 13. Infrared thermography was added in 1992 as an emergent technique; it was fully recognized only in 2007 when the ASTM E2582 concerning application of flash thermography for inspection of aerospace composite panels was released [13]. The NDT methods, actually, covered by ISO 9712 are:

Acoustic emission testing
Eddy current testing
Infrared thermographic testing
Leak testing
Magnetic testing
Penetrant testing
Radiographic testing
Strain gauge testing
Ultrasonic testing
Visual testing

2.5 BASIC PRINCIPLES OF SOME NONDESTRUCTIVE TESTING TECHNIQUES

In this section some of the techniques that could be used for NDT will be reviewed. In particular, for several of them only general descriptions will be given, while ultrasonics (which is the most widely used, also for the inspection of composites in the aerospace sector) will be described in greater detail. Nondestructive testing with infrared thermography, which falls within the main topic of this book, will be discussed in the next chapter.

2.5.1 Visual Testing

Visual testing (VT), or visual inspection, represents the first step of NDT examination. In fact, it has several real advantages: simplicity, rapidity, low cost, minimal training and equipment requirements, and the possibility to be performed while the part is being used or processed. A variety of product forms, including castings, forgings, machined components and weldments can be inspected. It is generally based on visual inspection by a human being. The imperative requirement is, of course, the visual acuity of the inspector [14,15] and moderate experience as well. Mechanical (such as templates, moulds etc.) or optical (at least a magnifying lens (Fig. 2.4)) aids may be necessary to perform VT. Basically, VT requires:

- cleanup of the part surface,
- adequate illumination of this surface,
- line-of-sight contact with the portion of the part to be inspected,
- thorough understanding of the nature and origin of discontinuities and potential defects within the test object,
- familiarity with the specifications controlling the inspection process.

Often, visual inspection is self-consistent, eliminating the need for further testing and the associated costs. Sometimes it locates areas that require subsequent inspection by other NDT techniques.

Figure 2.4 Visual testing aid.

2.5.2 Acoustic Emission

The basic principle of acoustic emission refers to the generation of transient elastic waves within a material produced by a sudden redistribution of stresses in it [16,17]. More specifically, when a structure is subjected to an external stimulus (change in pressure, load or temperature), localized sources may trigger the release of energy in the form of stress waves that propagate to the surface and can be recorded by appropriate sensors. Such sensors, generally piezoelectric ones, are placed on the material surface, directly attached to it, in key points. Detection and analysis of emitted acoustic signals can supply valuable information regarding the origin and importance of a discontinuity in a material. A sketch showing the position of some piezoelectric sensors and their working scheme is given in Fig. 2.5.

Acoustic emission testing (AT) is a passive method because practically no energy is introduced into the object under examination, except for the external stimulus; instead, it simply listens to the energy released by the object. Then, AT tests are often performed on structures while in operation, as this provides adequate loading for propagating defects and triggering acoustic emissions. Owing to its inherent features, AT is classified as a health monitoring technique rather than NDT one. It is used in many industrial

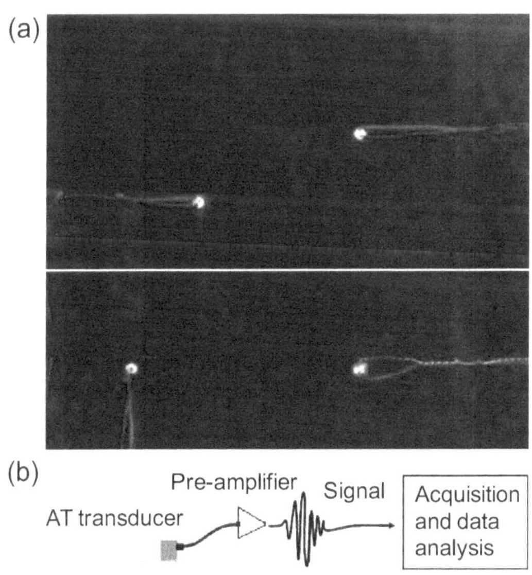

Figure 2.5 Sketch of the acoustic emission approach. (a) Position of AT transducers on a carbon fibre-reinforced polymer panel and (b) AT working scheme.

applications, for assessing structural integrity, detecting flaws, testing for leaks or monitoring weld quality, and is extensively used as a research tool.

However, AT suffers from loud service environments that contribute extraneous noise to the recorded signals, making it crucial to have signal discrimination for successful application. AT systems can only qualitatively gauge how much damage is contained in a structure. Then, to quantitatively evaluate size and depth of buried defects for the overall acceptability of a part, other NDT methods are necessary.

2.5.3 Eddy Currents

ET uses electromagnetic induction to detect defects in electrically conductive materials. The basic principle relates to the variable magnetic field created by an alternating current (AC) running through a coil as depicted in Fig. 2.6. When the AC flows in the coil, which is in close proximity to a conducting surface, the magnetic field of the coil will induce a circulating (eddy) current nearby that surface. The magnitude and phase of the induced eddy current will affect the loading on the coil and thus its impedance [18]. Then, these eddy currents may be detected either with a second *receiver* magnetic field sensitive sensor or by measuring the changes to the current flowing in the primary *excitation* coil. Variations in the electrical conductivity or magnetic permeability of the test object, or the presence of any flaws, will cause a change in the eddy current and a corresponding change in the phase and amplitude of the measured current. The described mechanism represents the basis for eddy current inspection.

ET can detect very small cracks in, or near, the surface of the material, also in the presence of physically complex geometries (by choosing the most adequate probe) with minimal preparation of the surface. It is also useful for making electrical conductivity and coating thickness

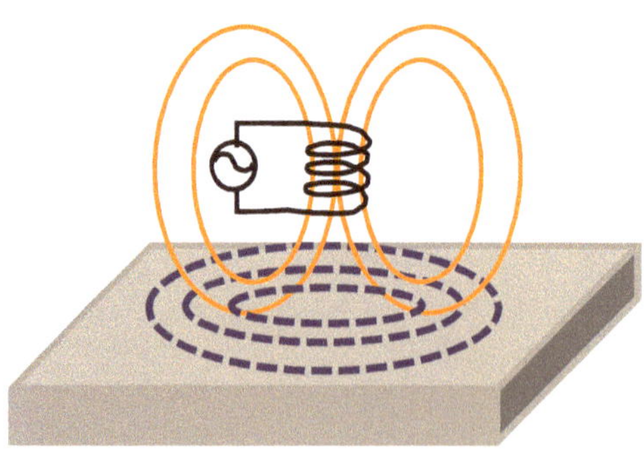

Figure 2.6 Sketch of the induction principle.

measurements. However, there are several limitations in the use of eddy current inspection. As a main disadvantage, the technique is susceptible to many effects that may compromise its response, making discrimination of the defects difficult. In fact, the response from an eddy current probe may come not only from a flaw but also from many other parameters. Thus, the basis of much of the technology of eddy current inspection consists of eliminating any undesired effects from results. The main factors to be considered for assessing eddy current measurements are:

Material electrical conductivity: it has a very direct effect on the eddy current flow: the greater the conductivity of the material, the greater the flow of eddy currents nearby the surface. The electrical conductivity of a material may be affected by different factors, such as material composition, heat treatment, mechanical hardening etc.

Material magnetic permeability: indicates the aptitude of a material to become magnetized. As a general indication, the permeability of nonferrous metals (eg, copper, brass, aluminium etc.) and of austenitic stainless steels is practically the same as that in vacuum. Instead, that of ferrous metals is much higher and has a very significant influence on the eddy current response; in addition, it is not uncommon for the permeability to vary greatly within a metal part due to localized stresses, heating effects etc.

Geometry: geometrical factors, such as curvature, edges, grooves etc., affect the eddy current response. This should be taken into account when choosing the testing probe. In fact, eddy current probes can be made helicoidally or on a planar disposition. Planar probes provide a better contact with the material and can be flexible as well, allowing the inspection of complex geometry surfaces. Another problem affecting the eddy current response may be the material thickness if it is less than the effective depth of penetration.

Proximity/Liftoff: the distance between coil and material is of great concern since the closer a probe coil is to the surface, the greater will be the effect on that coil. This has two main consequences: a 'liftoff' signal as the probe is moved on and off the surface and a reduction in sensitivity as the distance between coil and surface increases.

In addition, the eddy current response is greatly affected by the chosen test frequency, even if this is not a great problem since it can be quite easily controlled. With regard to the depth of penetration, it has to be considered that the strength of the response is greatest on the surface of the metal being tested, while it declines with depth. It is physically convenient to define the *standard depth of penetration* where the eddy current is $1/e$ (37%) of its surface value. As a final point, flaws lying parallel to the probe may be undetectable. Generally, eddy currents are mostly used to detect cracks in metallic parts. In some applications, they can be used in conjunction with infrared thermography, but this will be discussed in the next chapter. However, only some basic information has been reported here; for further information on ET, please see specifically related literature [19,20].

2.5.4 Penetrant Testing

The inspection with penetrant liquids, also referred to as dye penetrant inspection, or simply called penetrant testing, is a NDT method widely used to locate surface-breaking defects in all nonporous materials (metals, plastics or ceramics). Basically, a low surface tension fluid, which may or may not be fluorescent, depending on the type of dye used, is applied on the surface of the tested component by dipping, spraying or brushing it. After an appropriate period of time, called *dwell time*, which is required for the liquid to penetrate into the flaw cavities, the excess penetrant is removed from the surface and a developer is deposited on the latter. This yields the formation of indicator patterns in the layer where flaws are located [21]. The developer acts as a blotter and helps to draw penetrant out of the flaw so that an invisible indication becomes visible to the inspector.

Penetrants with an added visible dye (colour contrast) are typically red and require sufficient *white light* to ensure proper inspection. Penetrants with an added fluorescent dye need to be used in darkened conditions with an ultraviolet, or *black light*, because the excited dyes emit brilliant yellow-green light that contrasts strongly against the dark background. This method is quite simple and portable and is used to inspect a variety of product forms including castings, forgings and weldments. The structural steel, automotive, petrochemical, power generation and aerospace industries are a few examples of those that utilize liquid penetrant inspection.

However, against the method simplicity, many aspects require great attention. These include the environment under which the test is performed, part surface finish and size of defects sought. As a primary prerequisite, the test surface must be cleaned to remove any dirt, paint, oil, grease or any loose scale that could either keep penetrant out of a defect or cause irrelevant or false warnings. One must also ensure that the test chemicals are compatible with the surface material to prevent the examination from causing permanent staining or degradation. As a further requirement, the test surface needs to be cleaned after inspection to assure a complete removal of chemicals that may be detrimental to the part life if it remains entrapped inside cracks. In this context, small cracks may be critical since they may be open enough to suck penetrants and get soaked, but they may be narrow enough to prevent the fluid from being rejected. To sum up all the steps in the inspection with penetrants, see the workflow shown in Fig. 2.7.

2.5.5 Radiographic Testing

When we pronounce the word *radiography*, our gaze goes suddenly to the safety sign seen in the hospital or in the dentist's studio (Fig. 2.8), and we instinctively feel a sensation of danger.

RT is perhaps the most known method since it is used in different fields involving the medicine and engineering. It is an NDT method that exploits the capability of short wavelength electromagnetic (high-energy) photons to penetrate various materials. In fact, this method uses X-ray (Fig. 2.9) to produce a radiograph of the part under exam, showing any changes in thickness, defects (internal and external) and assembly details. Also, gamma rays may be used for specific applications, even if X-rays are mostly used and the radiographic method is generally referred to as X-rays.

RT basically consists of sending X-rays through the material under inspection and measuring the attenuation of the exiting ray intensity. The result is a black/white picture in which anomalies become visible as local

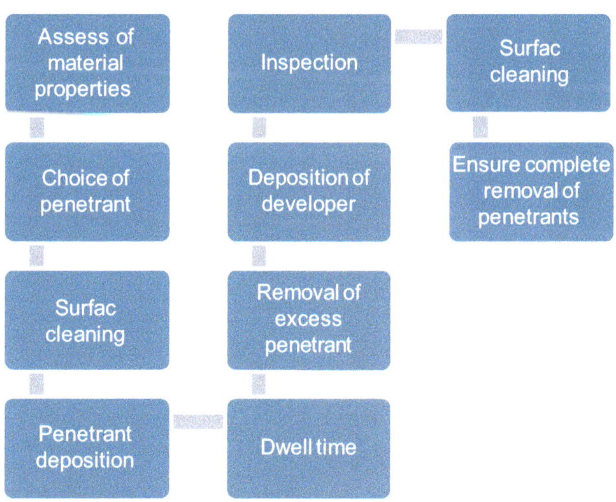

Figure 2.7 Penetrant inspection – complete workflow.

Figure 2.8 Radiographic testing – safety sign.

change of luminous intensity [22]; in fact, radiation decays in different ways through different materials and thicknesses (ie, the light/shade effect strongly depends on the local radiometric density and thickness of the material). Thus, a detailed profile of the internal defect distribution can be obtained [23]. A broad range of material thicknesses can be inspected, as well as complex shapes. Radiography for the carbon fibre-reinforced polymer inspection is principally used to detect subsurface imperfections or discontinuities such as porosity, inclusions or fibre waviness [24]. Radiography has numerous advantages since X-rays can pass through relatively thick objects and visualize the inside of visually opaque objects. However, it is quite expensive and involves health risks [25].

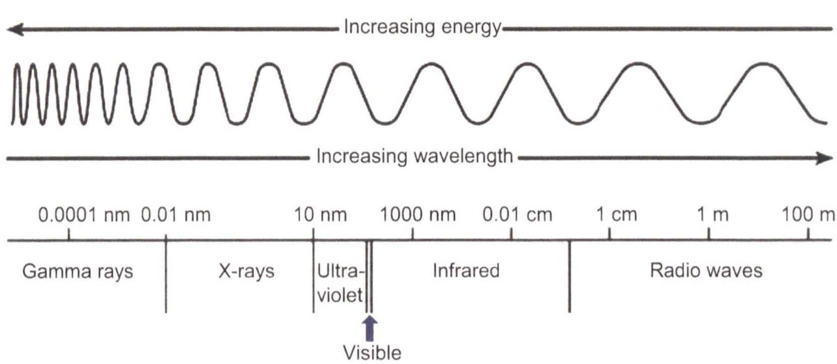

Figure 2.9 The electromagnetic spectrum: position of X- and gamma rays (wave not to scale).

The radiographic system includes an X-ray source (X-ray tube) and an image detection system, which may be either a film support or a digital capture system (*computed*, or *digital*, radiography). X-ray photons are emitted from a target struck by an electron beam at high velocity. Such a target forms the anode of the vacuum tube, also called the thermionic tube. Electrons are produced by a hot filament (commonly tungsten), which is heated to incandescence. The energy of the radiation depends on the voltage applied to the anode (to attract the negatively charged electrons). More specifically, an increase in voltage to the anode causes an increase in the velocity of electrons with, in turn, an increase in the radiation energy. Since the higher the energy the shorter the wavelength, increasing the voltage would result in the production of X-rays of greater penetration ability.

X-ray photons are emitted by the target in all direction, with the highest intensity in between 60 and 90 degrees from the beam. They carry enough energy to ionize atoms and disrupt molecular bonds, and therefore are harmful to living tissue. A very high radiation dose over a short amount of time causes radiation sickness, while lower doses can give an increased risk of radiation-induced cancer.

In addition, the inspection of complex geometrical objects requires more than one angle of view, which lengthens the complete testing, also because of the time required for film developing in nondigital equipment. In principle, the process could be shortened with the use of computed radiography. This offers the possibility to examine different thicknesses by simply adjusting the image brightness, and reduces the developing time needed for film X-rays; besides, there is the advantage offered by digital images to be enhanced and analyzed via *software*. Unfortunately, this methodology still has a much lower resolution that makes it unsuitable for some specific requirements.

2.5.6 Ultrasonic Testing

UT is based on the principle that an ultrasonic wave, of frequency higher than 20 kHz (above human hearing range), modifies passing through a material. Ultrasonic waves behave similarly to light waves, in that they can be reflected, refracted and focused. In particular, when introduced into a homogeneous material, they travel in a straight line and at constant speed until they encounter a discontinuity. Then, waves undergo amplitude reduction and reflection at interfaces between parts of different acoustic impedance [26].

The active element of most used transducers is a piezoelectric ceramic [27], which can be shaped in a particular way to produce different wave modes. Depending on the media and its geometry, ultrasonic waves can propagate in specific modes [28]; those mostly used for NDE are the so-called *bulk modes*, which can be of two types:

- *longitudinal*: the motion of particles is parallel to the direction of propagation, ie, the wave behaves as a compression and a consequent relaxation along their path;
- *transversal*: the motion of particles is orthogonal to the direction of propagation.

Both longitudinal and transversal modes are characterized by their specific propagation velocity, which depends on the material physical properties.

UT gives information only from/within the confines of the sound beam as it propagates into the tested material. Operators must use great caution when making inferences about the nature of the test material outside the limits of the sound beam. Material condition can significantly vary and results can be unreliable if operators do not exercise good judgment. Operators must be acquainted with the use of ultrasonic couplant, which is necessary between transducer and material surface. Testing abilities must be developed to use and apply couplants in a reliable manner to eliminate variations in couplant thickness, which can cause errors and inaccurate data. Calibration and testing should be performed under similar coupling conditions, with the minimum amount of couplant and constant pressure on the transducer.

2.5.6.1 Ultrasonic Testing Techniques

There are two main techniques to carry out UT: through transmission (TT) and pulse echo (PE).

In TT, two transducers are necessary, which are coupled at opposite sides of the inspected component. One transducer acts as a transmitter and the other one as a receiver (Fig. 2.10(a)). With this configuration, the presence of a discontinuity is discovered through the variation it induces to the amplitude of the received signal. TT is particularly suitable for the detection of porosity; in particular, it has been demonstrated that the slope of the attenuation curves, which depends on the frequency of the applied ultrasonic signal, can be directly related to the amount of porosity in the material under inspection [29]. This technique is also used for the estimation of the impact damage, to evaluate the extension of the damaged areas and to determine the fibres' orientation.

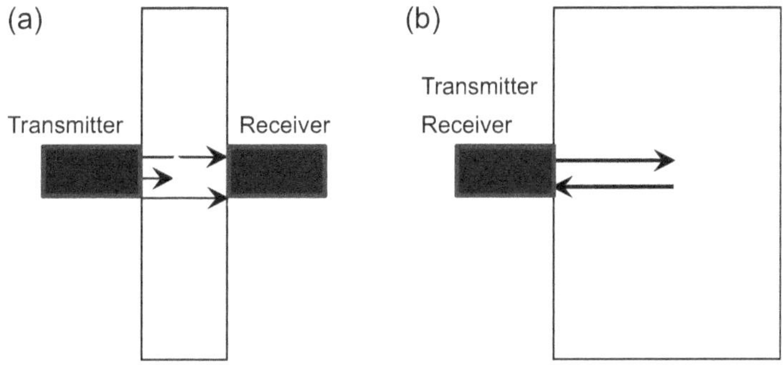

Figure 2.10 Ultrasonic configurations. (a) Through transmission and (b) Pulse echo.

For the PE (Fig. 2.10(b)), only one transducer is used, which alternatively works as transmitter and as receiver. In this way, information about the presence of flaws is provided by the amplitude and the time of flight of the reflected signal. To effectively carry out tests, the period of the excitation pulse of the unique transducer must be adequately long to avoid overlapping between transmitted and reflected waves. This method supplies information about the planar extension of the defect as well as its position through the thickness. PE technique is widely used for detection of porosity, fibre/matrix distribution [30], fibres' orientation [31], interlaminar debonding as well as impact damage [32].

When the couplant is water, another largely used technique is the double-trough transmission (DTT). It is a sort of PE that involves a transducer headed to one side of the sample material and a glass mirror put nearby its opposite side. Because of its very large impedance with respect to water, the glass mirror practically acts as a perfect reflector. In this way, the signal passes through water, the sample material, and then to water and vice versa. Since both the distance of the mirror from the sample surface and the ultrasound velocity in water are known, the attenuation due to the material can be measured. Moreover, using DTT, the sample thickness and the ultrasound wave velocity in it can also be measured as well as the sample porosity [33].

Both configurations, sketched in Fig. 2.10, refer to testing with a straight-beam transducer. For the PE, an important point to be taken into account is that the probability to detect a buried discontinuity depends on the amount of reflected sound received by the transducer. A volumetric discontinuity, such as a slag inclusion, which reflects the sound waves in different directions (Fig. 2.11), may be hardly detectable with a straight-beam transducer. Instead, an angle-beam (Fig. 2.12) transducer maybe more appropriate.

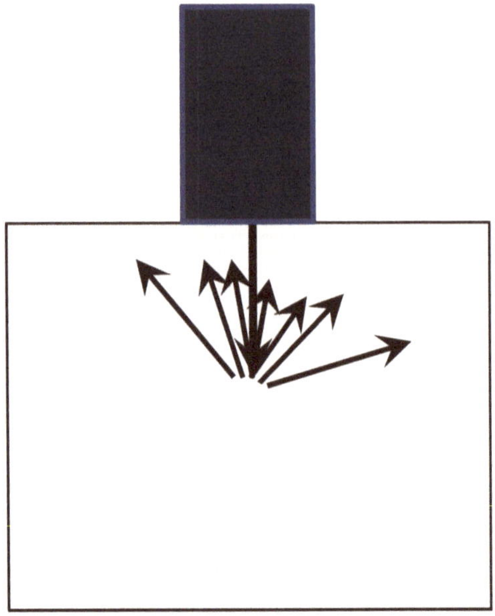

Figure 2.11 Reflections from a volumetric discontinuity.

Figure 2.12 Reflections from a volumetric discontinuity.

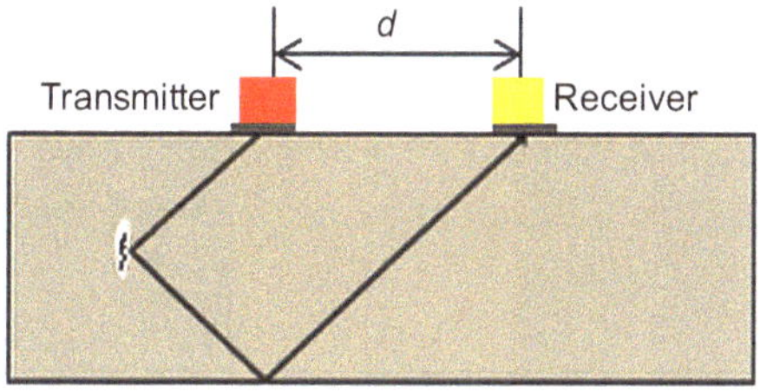

Figure 2.13 Tandem testing configuration.

However, other features may make defects undetectable. One example is the presence of a vertical discontinuity in a thick-walled test object; in this case the reflected sound waves do not return to the transducer. A solution may be the tandem technique [34], which involves the use of two transducers, one to act as transmitter and the other one as receiver, both positioned on the same side to perform tests in PE configuration (Fig. 2.13). Of course, this approach requires the two transducers to be moved over the surface of the test object and to be spaced apart at a certain distance *d*, which depends on the depth of the discontinuity. This concept has led to the phased array device [35]. Each transducer is much smaller than the conventional ones, and excitation can be provided to either separate elements or groups of them. This operational mode allows generation of wave fronts that can be directionally controlled and shaped by simply introducing a proper delay among elements. Generally, phased arrays are derived from a piezoelectric element that is divided into a number of segments. A wide range of sizes, shapes, frequencies and number of elements are available on the market.

Whatever the type of technique used, most important is the coupling of the transducer to the surface of the part under inspection. Due to the strong attenuation effect of air, to get useful levels of sound energy into the inspected material, great care must be paid to remove the air present between the transducer and the material under inspection. This can be achieved with two methods:

- Direct contact: special gels or oils are needed to assure the proper coupling with the inspected material.
- Coupling in water: all the transducers and the material under investigation are submerged in water (*immersion testing*). To carry out ultrasonic immersion testing, a water pool is required and, depending on the dimensions of the tested component, quite a large pool may be required. When a large pool cannot be realized, jet probes (*squirters*)

are used. The latter are transducers provided with a laminar water jet that impinges on the component surface, acting as a wave guide for the ultrasound beam. Water jet probes are widely used to carry out through-transmission testing, allowing for a relatively quick control of large components.

Regardless of the used technique, the received signals are turned into electrical signals by the transducers; thus, they can be amplified and displayed on an oscilloscope screen and/or acquired by suitable computer software that provides data visualization/representation.

2.5.6.2 Ultrasonic Testing Imaging Techniques

Data coming from UT can be presented in different formats; the following three are the most commonly used.

A-Scan

The amount of energy received by a transducer is displayed on an oscilloscope screen with the vertical axis representing amplitude and the horizontal axis representing time. In particular, for PE inspection, the A-scan allows positioning the transducer orthogonally with respect to the entry surface of the inspected material, with the consequence of a maximized correspondent echo amplitude (entry echo, EE). Moreover, choosing a transducer of proper frequency and diameter, the echo coming from the opposite side of the material is displayed, which is known as back-wall echo (BWE). When a defect is present, such as a delamination, the echo, which is due to the interface between defect and sound material, is seen between the EE and the BWE. Depending on size and nature of the defect, the amplitude of BWE decreases as far as it disappears [36].

If the ultrasound velocity in the material is known, by measuring the time spent by the signal to reach the defect and then come back to the transducer, which is called time of flight, the depth of the defect can be evaluated. In the A-scan imaging, the axial resolution depends on the wavelength and on the pulse duration. The spatial resolution depends on the distance between the transducer and the sample surface; therefore, it depends on the divergence of the beam and on the ratio between the transducer diameter and the wavelength.

B-Scan

The B-scan presentation is a profile (cross-sectional) view of the test specimen. In other words, it includes A-scans along a line, which may be obtained by the transducer moved with a computed-traversing system.

Measurements can be graphically represented with the transducer position on the x-axis and the time on the y-axis. To each point of the obtained graph is associated a signal intensity, which is attenuated by reflections and/or scattering phenomena, occurring nearby that point. Thanks to the B-scan, not only the position of a defect but also its extension along the scanning line is provided. The axial resolution is the same as for the A-scan, but the spatial resolution depends also on the scanning step (ie, the distance between two acquired points).

C-Scan

The C-scan is probably the most useful imaging technique since it supplies details about the planar extension of defects through their 2D-projected image. In this case, the transducer is required to move along a pattern covering the entire sample surface. When PE is used, a time-of-flight C-scan can also be obtained, providing the position through the thickness of the defects. Both axial and spatial resolution are the same as for the B-scan imaging.

2.5.6.3 Pros and Cons of Ultrasonic Testing

UT has many advantages:

1. high depth of penetration
2. single-sided access
3. high accuracy in determining the depth of internal flaws and the thickness of parts with parallel surfaces
4. nonharmful to nearby personnel and other equipment
5. portability

But UT also has some limitations:

1. the surface must be accessible
2. a coupling medium is necessary
3. calibration with reference blocks is required
4. careful attention by experienced technicians is required
5. inspection of parts that are rough, inhomogeneous, very small or thin is crucial
6. the surface requires cleaning up
7. inspected items must be water resistant, when coupling in water

2.5.6.4 Toward Advanced Ultrasonic Techniques

The main limitation of ultrasonic inspection is the presence of the coupling medium, which can contaminate the tested parts (especially aerospace components). Thus, advancements in ultrasonic technology are mainly geared toward the possibility to eliminate the coupling medium and to produce ultrasonic waves in a different way.

Noncontact Ultrasonic Inspection

Noncontact of course means coupling in air. Recently great efforts have been devoted to the development of probes working with air coupling. The main problem resides in the strong attenuation induced by air. In fact, in the presence of a gap of air between the transducer and the object surface, a great amount of ultrasonic energy is lost during transmission and, the higher the frequency, the higher the attenuation. The attenuation effects become less important when the through transmission set-up is preferred and for low-frequency (30 Hz up to 100 Hz) or for medium-frequency (0.5 MHz up to 3 MHz), applications. This approach may be useful for the inspection of honeycomb-sandwich composites [37]. Low-frequency transducers provide a better signal-to-noise ratio; but, the medium-frequency transducers are able to perform inspection of composite materials. To reduce the acoustic mismatch between air and material, capacity ultrasonic transducers have been conceived [38].

Novel Methods

The inspection by conventional ultrasonic methods is based on the capability of what are known as listening technologies. However, ultrasonic waves, when propagating inside a medium, generate complex waveforms that may lead to signal misinterpretation with either overestimation or underestimation of defects. Therefore, attempts are underway to transform the classic listening approach into a seeing approach [39].

One novel approach is laser ultrasonics in which laser light is used to induce ultrasound by either ablating the medium or causing a sudden thermal expansion. The resulting ultrasonic wave packets can be detected by a transducer, or by interferometric techniques, taking advantage of the superficial deformation induced by the incoming ultrasonic wave. The laser beam provides noncontact scanning that allows application on complicated shape objects. The effectiveness of this technique has been proven mostly for metals but it is also used for composites.

A laser can also be used to induce Lamb waves, which are particularly useful in the inspection of thin structures [40]. The lasers most commonly used for ultrasound generation are solid-state Q-switched Nd:YAG and gas lasers (CO_2, or excimers) because of their short pulse (from tens of nano-seconds to femtoseconds) and high-peak power. Detection lasers are continuous or long pulse (typically tens of microseconds) of long-coherence length. For components with local thickness variations, and/or non-homogeneous fibre-volume fraction, it has been demonstrated that the use of Lamb waves could lead to clearer defects detection [41] over classical UT performed with bulk waves as preferred modes, especially in the presence of diluted porosity [42].

However, these types of waves are generally too complex to be used in NDE due to difficulties associated with their dispersive nature (the wave velocity is frequency dependent, instead of what is seen for the bulk waves) and the existence of multiple modes propagating simultaneously. This requires specific preliminary studies in order to select the most suitable modes to be used and to gain knowledge about their behaviour when they interact with defects [43].

In recent years, the increasing interest toward guided waves is related to the possibility of creating sensors embedded within the airplane structures to be inspected. This leads to the possibility of obtaining online information about the structural integrity (structural health monitoring), avoiding wasting time to dismount the components for inspection. Different techniques can be used to get such smart structures [44] in which an array of sensors is attached or embedded within the structure. The position of the sensors can be improved by creating a network of patches dialoguing with a control unit for the classification of the faults to obtain a rapid prognosis of the structural conditions. Moreover, a system capable of online communication also in flight can be obtained as suggested in the patent US7197931 [45].

2.6 NEW AND EMERGENT TECHNIQUES

Detection of subsurface defects in thin structures also can be effectively carried out with an optical technique such as shearography [46] or electronic-speckle-pattern interferometry [47]. Both are based on holographic interferometry, comparing the image taken when the material is under load with a reference image. The load is induced by thermal or mechanical methods like local under- or overpressure, or heating up the whole object. The loaded image is interferometrically superimposed to the reference image, which allows defects to be easily displayed. Defects such as disbonds, delamination, wrinkles, slag inclusions and out-of-plane deformation can be easily detected with an inspection rate of almost 1 m^2/min.

DISCLOSURE

Some figures, which refer to well known facts, were arranged traying inspiration from websites.

SUMMARY TO CHAPTER 2 AND INTRODUCTION TO CHAPTER 3

This chapter discussed NDT and evaluation of materials. The discussion started with the terminology, proceeding to some general requirements of non-destructive-testing techniques in terms of performance, portability and safety concerns. Then, a description of the actual standardized techniques, involving their basic principles, as well as their advantages and limitations, followed. A section was devoted to personnel qualifications and certification.

In particular, infrared thermography was not herein described since it represents the topic of this book and will be addressed in the following chapters, starting with its general basic principles in Chapter 3.

REFERENCES

[1] BS EN ISO/IEC 17020, Conformity Assessment – Requirements for the Operation of Various Types of Bodies Performing Inspection, 2012, p. 1.

[2] R. Marini, P. Ranos, Current Issues in Qualification and Certification of Non-destructive Testing Personnel in the Aerospace Industry, ECNDT, 2006. Th.3.6.5.

[3] (CEN) EN 4179:2000. Qualification and Approval of Personnel for Non-destructive Testing, 2000.

[4] (CEN) prEN 4179, Qualification and Approval of Personnel for Non-destructive Testing, Ed. P3, Feburary 2003.

[5] ICNDT Guide to Qualification and Certification of Personnel for NDT, International Committee for NDT, 2012.

[6] J. Thompson, Global review of qualification and certification of personnel for NDT and condition monitoring, in: 12th A-PCNDT 2006 – Asia-Pacific Conference on NDT. Auckland, New Zealand, 2006.

[7] Recommended Practice No. SNT-TC-1A, Personnel Qualification and Certification in Nondestructive Testing, 2006.

[8] ISO 9712, Non-destructive Testing – Qualification and Certification of NDT Personnel, 2012.

[9] ANSI/ASNT CP-106, ASNT Standard for Qualification and Certification of Nondestructive Testing Personnel, 2008.

[10] EN 473, Non-destructive Testing. Qualification and Certification of NDT Personnel. General principles, 2008.

[11] AIA-NAS-410. Aerospace Industries Association, National Aerospace Standard, NAS Certification and Qualification of Nondestructive Test Personnel, 2008.

[12] EN 4179, Aerospace Series. Qualification and Approval of Personnel for Non-destructive Testing, 2009.

[13] Standard Practice for Infrared Flash Thermography of Composite Panels and Repair Patches Used in Aerospace Applications, ASTM E2582−07, 2007.
[14] M.F. Deering, The Limits of Human Vision, web access February 2015: http://www.swift.ac.uk/about/files/vision.pdf.
[15] NDT Resource Center. Visual Acuity of the Human Eye, web access February 2015: https://www.nde-ed.org/EducationResources/CommunityCollege/PenetrantTest/Introduction/visualacuity.htm.
[16] A.G. Beattie, Acoustic Emission, Principles and Instrumentation JAE 2 96, 1983.
[17] Web access February 2015: https://www.nde-ed.org/EducationResources/CommunityCollege/Other%20Methods/AE/AE_Intro.htm.
[18] D.J. Hagemaier, Fundamentals of Eddy Current Testing, ASNT, 1990. ISBN: 0-931403-90-1.
[19] J. Blitz, Electrical and Magnetic Methods of Nondestructive Testing, CRC Press, 1991.
[20] C.J. Hellier, Eddy current testing, in: Handbook of Nondestructive Evaluation, The McGraw-Hill Companies, Inc., 2001 (Chapter 8).
[21] Y.A. Glazkov, Evaluation of material quality for liquid penetrant inspection based on the visibility of the indicator patterns of flaws, Russ. J. Nondestruct. Test. 48 (4) (2012) 208−217.
[22] R.M. Tapphorn, P. Spencer, H.D. Beeson, M. Kirsch, in: Nondestructive Testing and Evaluation, 11, Taylor & Francis, Milton Park, UK, 1994, pp. 139−147.
[23] C. Sauerwein, M. Simon, Inspection of fibre based technical composites by radioscopy and computed tomography, in: Proceedings of ROMA2000 15th WCNDT Rome, Italy, October 15−21, 2000, 2000, http://www.ndt.net.
[24] T.G. Marshall, Y. Hurmuzlu, Detection and quantification of fiber waviness in thick section composite components, in: Proceedings of AHS 55th International Annual Forum, Montreal, Canada, May 25−27, 1999, 1, 1999, pp. 818−824.
[25] International Atomic Energy Agency, Governmental, Legal and Regulatory Framework for Safety, IAEA Safety Standards Series No. GSR Part 1, IAEA, Vienna, 2010.
[26] Standard Practice for Ultrasonic Testing of Flat Panel Composites and Sandwich Core Materials Used in Aerospace Applications ASTM E2580-07, 2007.
[27] O.E. Mattiat, Ultrasonic Transducer Materials, Plenum Press, New York, 1971.
[28] J. Ktrautkrämer, H. Ktrautkrämer, Ultrasonic Testing of Materials, fourth ed., Springer-Verlag, Heidelberg, 1990.
[29] D.K. Hsu, Ultrasonic nondestructive evaluation of void content in CFRP, in: ANTEC '88 Proceedings of the 46th Annual Technical Conference Atlanta, USA, 1988, pp. 1273−1275.
[30] K.V. Steiner, in: J.E. Masters (Ed.), Damage Detection in Composite Materials, American Society for Testing and Materials (ASTM), Philadelphia, PA, 1992, pp. 72−84.
[31] J. Degriech, N.F. Declercq, O. Leroy, Ultrasonic polar scans as a possible means of nondestructive testing and characterization of composite plates, Insight 45 (3) (2003) 196−201.
[32] R.A. Smith, A.B. Marriott, L.D. Jones, Delamination sizing in fibre-reinforced plastics using pulse-echo amplitude, Insight 39 (5) (1997) 330−336.
[33] P.W. Lorraine, R.J. Filkins, M. Dubois, T.E. Drake Jr., J.B. Deaton Jr., System and Method of Determining Porosity in Composite Materials Using Ultrasounds, US6684701, 2004.
[34] NDT.net, September 2000, vol. 5 (9). http://www.ndt.net/article/v05n09/berke/berke1.htm, accessed on September 2015.
[35] G.E. Georgeson, Array-Based System and Method for Inspecting a Workpiece with Backscattered Ultrasonic Signals, US7712369, 2010.

[36] C. Meola, C. Toscano, Nondestructive evaluation of carbon fiber reinforced polymers with ultrasonics and infrared thermography: an overview on historical steps and patents, Recent Pat. Mater. Sci. 5 (1) (2012) 48–67.
[37] L.E. Hampton, Non-contact Ultrasonic Testing Method and Device for Ceramic Honeycomb Structures, US0266789, 2007.
[38] J. Song, D.E. Chimenti, Method and Apparatus for Air-Coupled Transducer, US7759842, 2010.
[39] J. Takatsubo, H. Miyauchi, H. Tsuda, N. Toyama, K. Urabe, B. Wang, Generation laser scanning method for visualizing ultrasonic waves propagating on a 3-D object, in: Proc. of the 1st International Symposium on Laser Ultrasonics: Science, Technology and Applications July 16–18, 2008, Montreal, Canada, 2008.
[40] T.W. Murray, C. Prada, O. Balogun, Non-destructive Imaging, Characterization or Measurement of Thin Items Using Laser-Generated Lamb Waves, US7798000, 2010.
[41] D.E. Chimenti, R.W. Martin, Nondestructive evaluation of composite laminates by leaky Lamb waves, Ultrasonics 29 (1991) 13–21.
[42] D.E. Chimenti, Y. Bar-Cohen, Properties of Composite Laminates Using Leaky Lamb Waves, US4674334, 1987.
[43] K. Maslov, T. Kundu, Selection of lamb modes for detecting internal defects in composite laminates, Ultrasonics 35 (1997) 141–150.
[44] V. Giurgiutiu, J. Bao, A.N. Zagrai, Structural Health Monitoring System Utilizing Guided Lamb Waves Embedded Ultrasonic Structural Radar, US6996480, 2003.
[45] H.Y. Kim, Sensors for Monitoring Structural Health Conditions, US7197931, 2007.
[46] Standard Practice for Shearography of Polymer Matrix Composites, Sandwich Core Materials and Filament-Wound Pressure Vessels in Aerospace Applications ASTM E2581, 2007.
[47] W. Weikl, D. Findeis, E. Schnack, J. Gryzagoridis, Comparing optical interference techniques for the nondestructive detection of delaminations in layered composites, in: Proceedings of the 7th International Conference on Composites Engineering, ICCE7, Denver, Colorado, July 2–8, 2000, 2000, pp. 921–922.

CHAPTER 3

Infrared Thermography Basics

3.1 DEFINITIONS AND APPLICATIONS

First of all, we need to somehow define *infrared thermography (IRT)*. In the literature we can find terms like these: a tool, an instrument, a method, a technique. Indeed IRT includes all of these but no one is able to depict a complete notion of it. Sometimes it is referred to as an instrument, being confused with the infrared (IR) imaging device. Often it is identified with the shorter term – thermography; this is perhaps because of the speedy goings in our society dominated by contractions, to accomplish with SMS and email communications. However, from a rigorous point of view the correct term is IRT, since thermography alone would mainly refer to a map (a distribution of temperature signatures), without saying anything about the way it has been obtained. In fact, a temperature map can also be obtained by using liquid crystals or something else.

IRT is a discipline that relies on physical principles, a theoretical basis and practical approaches. Practically, it includes an IR detecting device to sense the thermal energy that is radiated from objects in the IR band; such energy, with the aid of specific software and basic relationships, is transformed into a video signal and, finally, into the object surface temperature map. Naturally, this is a simplified description; in reality, the way the final temperature map is accomplished is a rather complex procedure that involves many others disciplines like electromagnetism, electronics and signal treatments, heat transfer, etc.

IRT is being used in a broad number of application fields and for many different purposes; indeed, any process that is temperature dependent may benefit from the use of an IR device. In other words, an IR imaging device should be considered as a precious ally to consult for diagnostics and preventative purposes, for understanding of complex fluid dynamics phenomena, or for material characterization and procedures assessment, which can help improve design and fabrication of products. IRT may accompany the entire life of a product, since it may be used to control the manufacturing process (online process control), to nondestructively assess the final product integrity and to monitor the component in service. Of course, for the success of any

Infrared Thermography in the Evaluation of Aerospace Composite Materials
ISBN 978-1-78242-171-9
http://dx.doi.org/10.1016/B978-1-78242-171-9.00003-6

application it is most important to choose the most adequate IR camera and test procedure, as well image processing and data analysis for the specific use required.

Some important information on basic theory, detector types and actually available IR devices will be given in the following discussion. In addition, the reader is invited to be up-to-date with the proceedings of the main conferences concerned with IRT, including the Quantitative InfraRed Thermography (QIRT; biennial, starting from 1992), Advanced Infrared Technology and Applications (AITA; biennial, starting from 1991), SPIE Thermosense Conferences (annual, starting from 1978) and Inframation (from 1999).

3.2 THE BASIC RELATIONSHIPS AND HISTORICAL STEPS OF INFRARED TECHNOLOGY

The origin of IR technology began with the discovery by Sir William Herschel of thermal radiation, outside the deep red of the visible spectrum; this was first called invisible light and later *infrared*. From that, a century went by with a proliferation of ideas and debates between eminent scientists (amongst them, Macedonio Melloni, Gustav Kirchhoff, Clark Maxwell, Joseph Stefan, Ludwig Boltzmann) until the early 20th century when the German physicist Max Planck derived the basic law of radiation, which precisely describes the spectral distribution of the radiation from a blackbody (Planck's law):

$$E_{\lambda\mathrm{b}} = \frac{2\pi \mathrm{hc}^2}{\lambda^5\left(e^{\mathrm{hc}/\lambda \mathrm{k_b} T} - 1\right)} \qquad [3.1]$$

where: $E_{\lambda\mathrm{b}}$ is the blackbody monochromatic (spectral) radiation intensity, h is *Planck's constant* ($\mathrm{h} = 6.6 \times 10^{-34}$ J s), c is the speed of light ($\mathrm{c} \cong 3 \times 10^8$ m/s), λ is the wavelength of the radiation being considered, $\mathrm{k_b}$ is *Boltzmann's constant* ($\mathrm{k_b} = 1.38 \times 10^{-23}$ J/K) and T is the absolute temperature of the blackbody (K). Planck's law is plotted in Fig. 3.1 for several absolute temperature values in the range 200–6000K; a family of curves is obtained. For each curve, $E_{\lambda\mathrm{b}}$ starts from zero for $\lambda = 0$, then increases rapidly up to a maximum and, finally, decreases toward zero again at very long wavelength values. As can be seen, the higher the temperature, the shorter the wavelength at which the maximum occurs.

Later, Albert Einstein exploited Planck's idea suggesting that an electromagnetic wave, such as light, could be described by particles called *photons* with discrete energy dependent on their frequency. This led to the

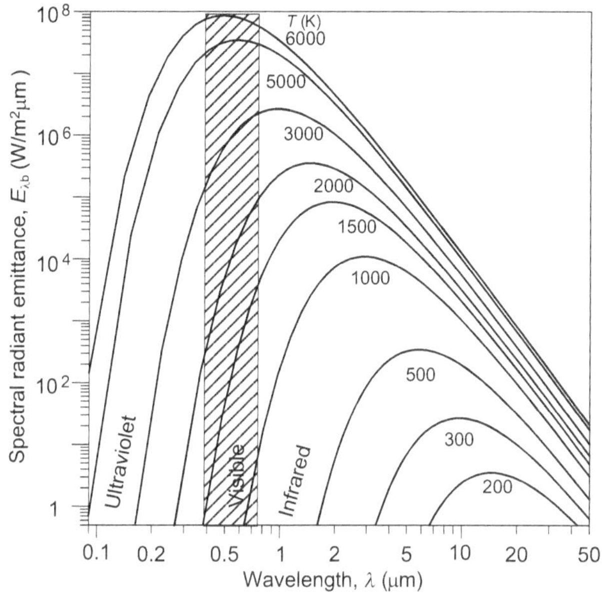

Figure 3.1 Blackbody emissive power against wavelength (Planck's law).

quantum description of electromagnetism [1–3]. Thermal energy is emitted in the form of discrete quanta, or photons. Each photon has energy E_{ph} equal to its frequency f_{ph} multiplied by the Planck's constant h, and since, according to Einstein, photons move at the speed of light c, the following relationship can be written:

$$E_{ph} = hf_{ph} = h\frac{c}{\lambda} \qquad [3.2]$$

which states that the energy is inversely proportional to the wavelength, ie, the higher the energy, the shorter the wavelength. The electromagnetic spectrum is shown in Fig. 3.2; more specifically, from the figure it is possible to see the location of the IR band in terms of wavelength and frequency (the frequency increases for decreasing the wavelength). The same figure also depicts the extension of the thermal region, which embraces the entire IR and visible bands, as well as the upper part of the ultraviolet band.

By differentiating Planck's law Eq. [3.1] with respect to λ and by looking for the wavelength λ_{max}, which emits the maximum radiation intensity, the *Wien's displacement law* (Wilhelm Wien) is obtained:

$$\lambda_{max} = \frac{d_w}{T} \qquad [3.3]$$

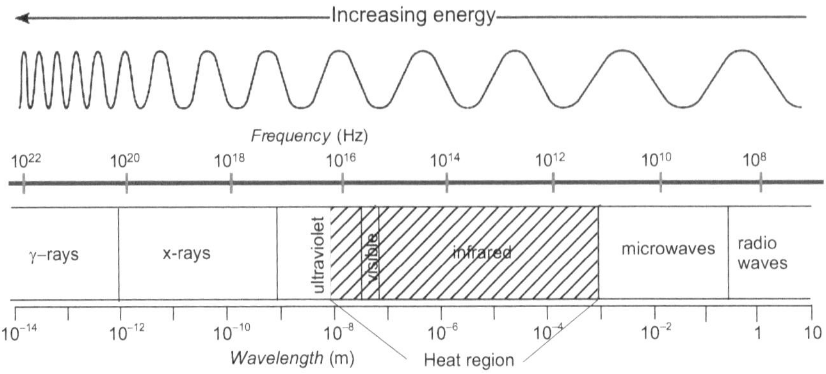

Figure 3.2 Representation of the electromagnetic spectrum (wave not to scale).

where d_w is called *Wien's displacement constant*, which is approximately equal to 2898 μm K. Then, Eq. [3.3] practically states that at ambient temperature (about 300K) the radiation peak lies in the far IR at about 10 μm. The sun radiation (about 6000K) peaks at about 0.5 μm in the visible light spectrum (Fig. 3.1). Conversely, at the temperature of liquid nitrogen (77K) the maximum radiation intensity occurs at about 38 μm, in the extreme IR wavelength. Note that in Eq. [3.3], λ_{max} and T must be, respectively, given in micrometres and degrees Kelvin when using the reported value of Wien's constant. By integrating Planck's law over the entire spectrum ($\lambda = 0 \div \infty$), the total hemispherical radiation intensity (Stefan–Boltzmann's law) is obtained:

$$E_b = \sigma T^4 \qquad [3.4]$$

with σ the *Stefan–Boltzmann constant* ($\sigma = 5.67 \times 10^{-8}$ W/m^2 K^4).

It has to be pointed out that Eq. [3.1] describes the radiation emitted by a blackbody, which is the maximum value radiated by a body at a given temperature. Real objects almost never comply with the previous laws although they may approach the behaviour of a blackbody in certain spectral bands. A real object generally emits only a part E_λ of the radiation emitted by a blackbody at the same temperature and at the same wavelength $E_{\lambda b}$.

3.3 THERMAL RADIATION FROM A REAL OBJECT

Basically, every object, which is at a temperature above absolute zero, radiates a certain amount of energy due to the motion of the electrons of the substance it is made of. As anticipated in the previous section, the energy is emitted not as a continuous flow but in the form of discrete quanta, or photons. The magnitude of emitted radiation depends on the temperature and on the characteristics of the emitting object.

A real object, depending on both the bulk material it is made of and on the characteristics of its surface, may have a different ability to absorb, reflect and transmit energy. This ability is generally described through the relationship (energy conservation):

$$\alpha + \rho + \tau = 1 \qquad [3.5]$$

which links the *total absorptance* α to the *total reflectance* ρ and to the *total transmittance* τ. Eq. [3.5] can be also expressed on spectral basis as:

$$\alpha_\lambda + \rho_\lambda + \tau_\lambda = 1 \qquad [3.6]$$

relating to an elementary spectral interval, being α_λ the *spectral absorptance*, ρ_λ the *spectral reflectance* and τ_λ the *spectral transmittance*. However, scattering phenomena (Raman, Mie, etc.) may be also present; but they are herein neglected.

A surface may exhibit selective behaviour not only with respect to wavelength but also with respect to the direction of the propagating energy; more specifically, the energy, absorbed by a surface, comes not from the whole hemisphere but only from certain directions. Assuming a direction φ, Kirchoff's law states that, at a local thermodynamic equilibrium, the release of energy from a surface is the same as that absorbed from the same direction:

$$\alpha_{\lambda\varphi} = \varepsilon_{\lambda\varphi} \qquad [3.7]$$

This also means that the object remains at constant temperature with the rate at which it absorbs energy equal to the rate at which it emits energy; otherwise the object would warm, or cool, in contrast with the assertion of thermodynamic equilibrium.

The symbol ε indicates the emissivity, which is the fraction of the energy emitted by the black body: $E_\lambda = \varepsilon\, E_{\lambda b}$.

3.3.1 Objects' Characteristics

To allow for successful measurements with an IR imaging device, the measured object must be opaque (nontransparent) in the IR wavelength band and must have high emissivity.

One has to bear in mind that surfaces that are perfectly transparent do not exist, but they may be more or less transparent to radiation in certain wavelength bands. For example, glass is transparent to wavelengths within the visible range of the spectrum, but it generally absorbs radiation in the IR region. In fact, glass displays changeable behaviour with wavelength. Most solids are opaque to thermal radiation, so emission and absorption of radiation can be regarded as surface phenomena. Materials that are transparent in the IR band are used for the fabrication of IR optics (lenses and windows). However, as will be shown later, these materials are not transparent in the entire IR region but only in confined bands and are used for optics of systems working in such bands. IR windows are used in certain applications (eg, confined chambers, monitoring of a model inside a wind tunnel) that require an interface between the IR camera and the object under measurement. For this reason, users of IRT should be familiar with the material behaviour in the employed IR portion of the electromagnetic spectrum.

For an opaque material $\tau = 0$ and taking into account Eq. [3.7], Eq. [3.5] reverts to:

$$\varepsilon + \rho = 1 \quad [3.8]$$

which means that emissivity is the fundamental parameter to be known being:

$$\rho = 1 - \varepsilon \quad [3.9]$$

For practical use, materials are generally classified into two main categories: metals and nonmetals. Nonmetals are generally considered to have higher emissivity values with respect to metallic ones. However, bearing in mind that the IR camera views the surface of the object, what it is really important is the surface finishing rather than the material of which an object is composed. In fact, a metal object may display emissivity values in a broad range, practically from very low values (down to 0.01) for a sleek surface, up to very high values (close to 1) for a strongly oxidized, or sandblasted, surface.

According to Eq. [3.8], all real surfaces (except the blackbody, for which $\rho = 0$) reflect part of the incident radiation. More specifically, such a phenomenon consists of radiation coming from the surroundings that bounces off the target and is redirected. It is indifferently termed: reflectivity, reflectance or reflection. Herein, the term *reflectance* is used.

A surface may reflect the incident radiation in two modes: *specular* and *diffuse*.

Specular reflector — it is the surface for which all (or almost all) of the incident energy I_{inc} is thrown out of the surface as I_{out} in a single direction,

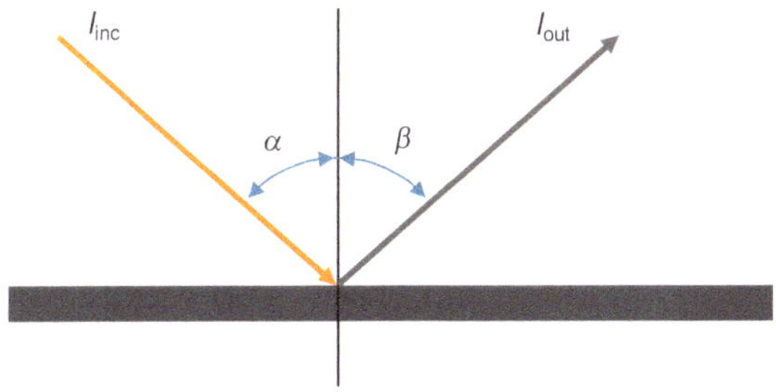

Figure 3.3 Specular surface.

with the incident angle α equal to the exit one β (Fig. 3.3). This happens when a surface is very smooth and highly polished (ie, a mirrorlike surface). For a perfect mirror: $\rho = 1$: $I_{inc} = I_{out}$.

Diffuse reflector – it is the surface for which the reflected energy leaves the surface almost uniformly in all directions (Fig. 3.4). This is the case of rough surfaces.

The ρ value of a surface depends on the surface roughness and the wavelength of the incident radiation. However, the previous examples are the extreme ends of the way a surface reflects energy; the behaviour of common objects lies between perfectly specular and perfectly diffuse reflectors.

For accurate temperature measurements with an IR system, the reflectance should be kept small ($\rho \to 0$). Otherwise, radiation emitted by the surroundings that strikes on the surface of the object of measure may be reflected, so being added to the radiation really emitted by it (Fig. 3.5).

Figure 3.4 Diffuse surface.

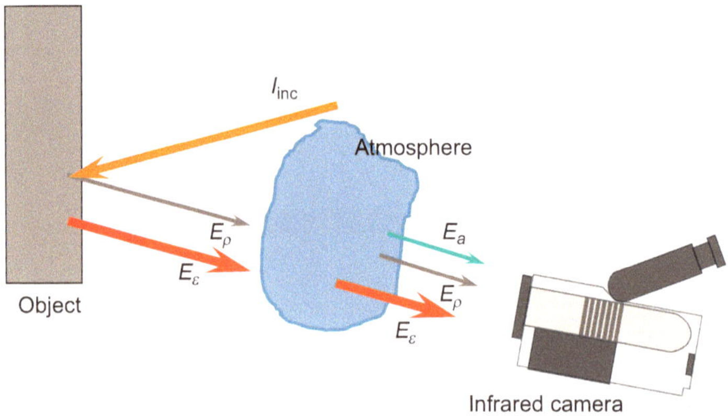

Figure 3.5 Components of energy reaching the infrared camera.

The total energy that departs from the surface E_d includes two contributions, E_ε and E_ρ:

$$E_d = E_\varepsilon + E_\rho \qquad [3.10]$$

where E_ε is the energy emitted by the surface owing to its temperature and emissivity values, or better:

$$E_\varepsilon = \varepsilon E_b \qquad [3.11]$$

with E_b coming from Eq. [3.1]. Instead, E_ρ is the reflected undesired energy, which does not depend on the measured object temperature.

Of course, the larger the ε, the smaller the ρ and consequently E_ρ, so, in turn, the temperature measurement is more accurate.

To perform measurements of temperature starting from thermal radiation, it is important to know exactly how much energy is emitted by the object and is captured by the IR detector. However, it is worth noting that E_ρ can be measured and accounted for. Even if depicted in Fig. 3.5, the procedure for this is not described herein because it is outside the scope of this work.

Apart from the surface finishing, its curvature deserves attention. In fact, only a blackbody emits radiation uniformly in all directions (diffuse). A real surface does not emit the same radiation in all directions; the maximum emission generally occurs in the direction normal to the radiating surface and decreases, becoming null for a direction parallel to that surface. A comparison between the directional emissivity of a blackbody and that of a nonmetallic, nonblack one is shown in Fig. 3.6. For the real body, a value of $\varepsilon = 0.7$ has been assumed in the normal direction (angle of observation

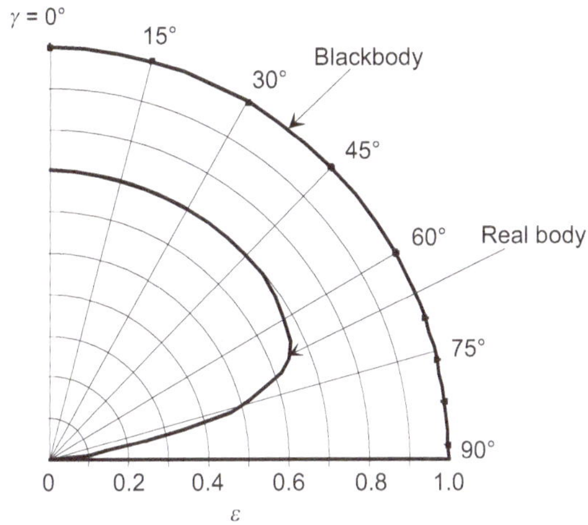

Figure 3.6 Directional emissivity for a blackbody and a real one.

$\gamma = 0$ degrees); this value remains essentially constant until γ about 50 degrees and then decreases first slowly and after that sharply toward zero for γ approaching 90 degrees (direction of observation parallel to the surface). Practically, Fig. 3.6 says that to measure the temperature of a curved surface, like the cylinder sketched in Fig. 3.7, the latter has to be subdivided into at least three sectors.

Going back to Fig. 3.5, before entering the detector, the radiation emitted by a body passes through the atmosphere and it may be affected by its absorption and emission, which must be taken into account. In a real

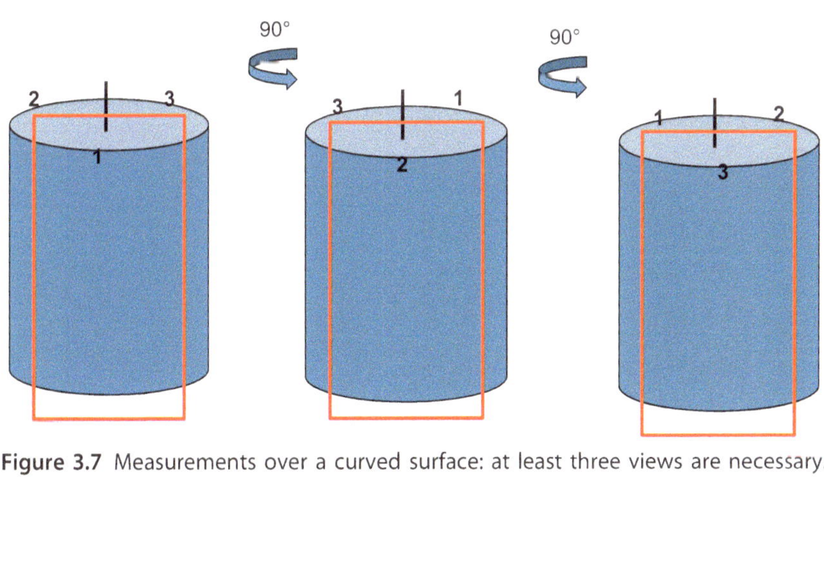

Figure 3.7 Measurements over a curved surface: at least three views are necessary.

situation, neglecting for now the absorptance of the interposed atmosphere, which will be considered later on, the total electromagnetic energy detected by the IR camera E_{tot} includes the energy emitted by the object E_ε, the reflected energy coming from the surroundings E_ρ and the energy emitted by the atmosphere E_a, so it can be written:

$$E_{tot} = AE_{b\varepsilon} + BE_{b\rho} + CE_{ba} \quad [3.12]$$

where the suffix b refers to blackbody emission and A, B and C are constants that account for the characteristics of the object surface and of the atmosphere between the object and the camera.

Normal atmosphere (eg, the air we breathe) consists of a mixture of gases. Generally, carbon dioxide absorbs IR radiation significantly in the 4.3 and 15 μm bands, the latter being the broad region of maximum intensity for the Planck function for emitters at temperatures between about 180 and 300K. Water vapour absorbs IR radiation in the 6.3 μm band and in several regions between 0.7 and 3 μm. Except for ozone, which has an absorption band in the 9.6 μm region, the atmosphere is relatively transparent from about 8 to 12 μm. The atmospheric transmittance is represented in Fig. 3.8 where the coloured area indicates the atmosphere transparency.

The presence of aerosols (salt particles, water droplets, dust, pollution haze) also contributes to atmospheric absorption. However, this contribution is generally negligible with respect to the molecular absorption. More than absorption, aerosols cause scattering (Mie scattering), ie, a redistribution of the radiation into all directions with loss in the wave travelling direction from source to detector. One may argue that also molecules cause scattering (Rayleigh scattering). Generally, molecular scattering can be

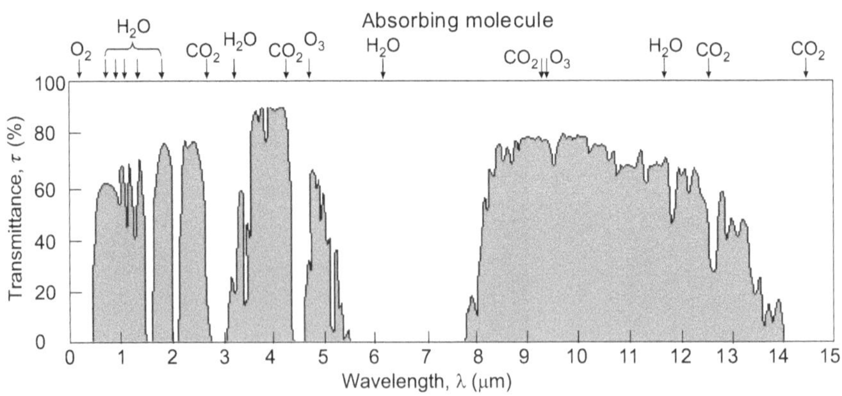

Figure 3.8 Atmosphere transmittance and gases that affect it.

considered negligible with respect to aerosol scattering. In any case, the scattering phenomenon is stronger in the visible region and in the IR region close to the visible one; as the wavelength increases the transmittance improves. In fact, IR cameras are also used to see through the fog or in the presence of a sandstorm.

Since Eq. [3.12], the atmosphere emits its own radiation. However, this contribution is important only in specific test conditions, ie, for large distances between the object and the IR camera and for measurements of a low-temperature object (close to the ambient temperature). The last phenomenon to be considered is turbulence, which may happen in the presence of strong wind or some convection transport effects. Turbulence can induce a random fluctuation of the refractive index, which may result in smearing of the generated image. This generally occurs for large-distance object detectors (order of hundreds metres) and with high sensitivity IR cameras.

The emission and attenuation induced by atmosphere could sometimes be a heavy burden for the operator since there are normally no easy ways to find accurate values of atmospheric emittance and transmittance for the actual case. An easy and convenient way is to account for the atmospheric interference during calibration of the IR system.

In some specific applications the IR camera views the object through a solid medium transparent to IR (eg, the window of a closed-circuit wind tunnel), and this poses the problem of design and material choice, plus an adequate calibration.

3.3.2 Emissivity Evaluation

The emissivity coefficient is commonly indicated with the symbol ε. In reality, the emissivity of real objects is generally wavelength dependent and indicated with the symbol ε_λ, which is called *spectral emissivity coefficient*. In fact, a real object emits only a part E_λ of the radiation emitted by a blackbody $E_{\lambda \mathrm{b}}$ at the same temperature and at the same wavelength:

$$\varepsilon_\lambda = \frac{E_\lambda}{E_{\lambda \mathrm{b}}} \qquad [3.13]$$

For some nonblackbody objects, the emissivity does not vary with the wavelength; these objects are called *greybodies*. A comparison between the spectral radiant emittance of a blackbody, a greybody and a real body, all at the same temperature, is shown in Fig. 3.9. From these plots, it is clear that

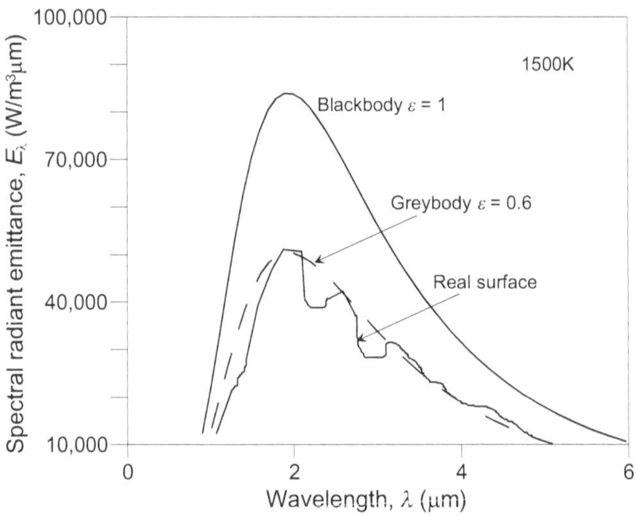

Figure 3.9 Emissive power for black, grey and real surfaces.

the radiation curve of a greybody is identical to that of a blackbody except that it is scaled down for the radiated power by the factor ε. Conversely, the radiation distribution of a real object depends on the wavelength.

The knowledge of the thermal emissivity is essential for accurate temperature measurements with an IR imaging system. The values of many of the most common materials are generously listed in the literature. However, these values are not at all useful when dealing with an IR system because of the following two main reasons.

The first one is that almost always we find tables with the parameter ε simply termed emissivity without any additional explanation about the direction and the wavelength that the information refers to.

Another problem arises regarding the state of the viewed surface, especially when dealing with emissivity values of metals. In fact, tables presented in literature include terms like polished, cast, rolled, oxidized, heavily oxidized etc., and emissivity values range inside very large intervals; of course, such values cannot be used to get accurate temperature measurements.

The better way is to directly measure the emissivity with the same IR camera according to the standards [4]. The procedure consists simply in comparing the radiation emitted by the material sample and that emitted by a blackbody at the same temperature. Almost every IR thermographic system is equipped with software for emissivity calculation.

For accurate thermographic measurements, it is preferable to work with high-emissivity surfaces. It is possible to increase the surface emissivity of highly polished metals, or reflectors, with deposition of thin films of paint or grease. This of course can be done in laboratory measurements and when the coating does not affect the surface performance. Almost every final produced part includes a protective coating of the surface; in most cases (except when specularity is a requisite), the choice of an opaque paint may fulfil both requirements: assure surface protection and allow for thermographic measurements. If a coating is not practicable, others expedients should be adopted such as a fine sandblasting the metallic surface. Then, it is quite easy to prevent undesired reflected radiation from hot surroundings by means of adequate screens.

3.4 INFRARED IMAGING DEVICES

IR systems are often classed as total-radiation radiometers and are considered to be based on the Stefan–Boltzmann law even if their detectors sense radiation in only a limited bandwidth of the IR spectrum. Therefore measurements made with IR radiometers are generally based on Planck's law. Actually, IR devices perform measurements in two main IR bands: *middle wave* (MWIR; 2–5 μm) and *long wave* (LWIR; 8–15 μm). Systems working in the near IR (NIR; 0.8–1.5 μm) window are tailored for specific applications. No measurements are practically possible in the region between 5.5 and 7.8 μm because the atmosphere is opaque to IR there (see Fig. 3.8).

The ultimate result of an IR system is a surface temperature map. However, up to a few years ago, the sensing element was zero-dimensional (a small area, practically a point). Image two-dimensionality was achieved by scanning mechanisms that consisted of oscillating mirrors, or rotating refractive elements (such as prisms), which allowed object scanning in both vertical and horizontal directions. Nowadays, the new systems are based on the staring focal plane array (FPA) technology.

The overall performance of an IR-imaging system is conventionally evaluated in terms of useful and accurate information that can be acquired per unit time. This can be expressed through several parameters such as: *thermal sensitivity*, or equivalent random noise level; *frame rate*, or number of images per unit time; *image resolution*, or number of independent measurement data points of which the image is composed; *intensity resolution*, or number of intensity levels, which allows for fine temperature differences, ie, dynamic range.

The core of the IR-thermographic system is the *IR detector* (or array of detectors) [5–11]. Basically, it is a transducer that mainly absorbs the IR energy emitted by the object (being measured) and converts it into a signal, usually an electrical voltage or current, which, as already said, can be transformed into a temperature map. Detectors used for IR technology can be grouped into two main categories: *thermal* and *quantum* (or photon) *detectors*.

3.4.1 Thermal Detectors

These detectors are also called energy detectors, or photon absorbers, because they absorb the incident energy and warm up; their temperature changes are measured through the variation of a temperature-dependent property of the material such as the electrical resistance. The main advantage is their response at room temperature and over a large band of the IR spectrum. The main disadvantage is their relatively slow response time (of the order of milliseconds), which makes them not suitable for high-frequency events. Examples of detectors that belong to this family are: thermopiles, bolometers, pyroelectric detectors and microcantilevers.

Thermopiles are based on the thermoelectric effect. A thermopile is composed of a number of thermocouples connected in series. As is well known, a thermocouple is obtained by joining two dissimilar metals. Heating the junction causes a voltage (Seebeck effect) that is proportional to the temperature change; thus, the temperature is computed from the measured voltage. The first thermopile was simply built using fine wire thermocouples (eg, copper–constantan). The temperature of the hot junctions increases until an equilibrium is reached between the rate of incident energy and the loss by conduction, convection and radiation to the surroundings. The thermoelectric output is measured by a suitable device and transformed into temperature (ie, the temperature of the target). The modern technology relays to thin films deposited on a substrate; this allows for reduced thermal inertia. The sensitivity can also be enhanced by encapsulating the thermopile in a low-thermal-conductivity medium (ie, to decrease losses to the surroundings).

Bolometers are based on the resistance change with temperature of a resistor element. Basically, it consists of two platinum strips, covered with lampblack, and one strip is shielded from the radiation and the other one exposed to it. The strips form two branches of a Wheatstone bridge; the resistance in the circuit varies when the strip, which is exposed to IR radiation, heats up and changes its electrical resistance. Nowadays, advances in silicon micromachining have led to the microbolometer technology, which includes a grid of vanadium oxide, or amorphous silicon heat sensors, atop a porous silicon bridge as thermal insulator and a mechanical supporting structure. The microbolometer grid is commonly found in different sizes: 160 × 120, 320 × 240, 640 × 512 and 1024 × 1024 arrays.

Pyroelectric detectors are based on the change of surface charge with temperature. In fact, the pyroelectric effect consists of a change of the crystal surface charge as a consequence of the change in the dipole moment when the crystal temperature is raised. For continuous operation, the common technique is to incorporate a chopping device inside the optical system and to create an AC output signal; the spurious signals caused by low-frequency ambient temperature can be filtered. Triglycine sulphate, lithium tantalate and polyvinyl fluoride are materials that exhibit the pyroelectric effects.

Microcantilevers are based on the dissimilar thermal expansion of substances (eg, silicon nitride and gold film). More specifically, a cantilever forms a capacitance in combination with a reference plate. The exposition to IR radiation causes temperature increase with consequent bending of the cantilever, which, in turn, alters the capacitance of the system. The intensity of IR radiation is measured as being proportional to the bending extension.

3.4.2 Photon Detectors

Photon detectors generate free electrical carriers in response to photon absorption. The main advantage is their very short response time (of the order of microseconds). A disadvantage is the need of cooling them down to cryogenic temperature to get rid of excessive dark current. In fact, electrons can be excited from the valence to the conduction band by photons having an energy E_{ph} that is larger than the energy gap E_g between the bands; there exists a wavelength λ_c named cut-off wavelength that corresponds to the energy E_g:

$$\lambda_c = \frac{hc}{E_g} \qquad [3.14]$$

beyond which no emission occurs. The value of E_g varies from semiconductor to semiconductor; it tends to increase as temperature increases (wavelength decrease). Therefore, to work at high wavelengths such as IR, the detector must be maintained at low temperature.

In the past, the detector was often located in the wall of a Dewar flask that was filled with liquid nitrogen, LN_2 ($\approx$ 77K). The new generation systems include miniature coolers based on the thermoelectric Peltier effect, or the Stirling cycle, and do not require any external cooling source. There are four main types of photon detectors: *photovoltaic (PV)*, *photoconductive*, *photo-emissive* and *quantum well IR photodetector* (QWIP). In particular, the PV and photoconductive types may be *intrinsic* or *extrinsic*.

PV intrinsic (electromotive force generation): The structure of a PV detector is based on a *p*—*n* junction device (two dissimilar materials), which, under IR radiation, generates photocurrents (current flows across the junction of the two materials). More specifically, when the detector is knocked by photons of energy greater than, or equal to, the energy band-gap, electrical carriers are swept across the photodiode and become separated by the potential barrier of the *p*—*n* junctions to create a current. PV devices operate in the diode's reverse bias region; this minimizes the current flow through the device, which in turn minimizes power dissipation. These detectors are generally fabricated from silicon (Si), germanium (Ge), gallium arsenide (GaAs), indium antimonide (InSb), indium gallium arsenide (InGaAs) and mercury cadmium telluride (HgCdTe) (also called MCT).

Photoconductive intrinsic (conductance change): In photoconductive (PC) detectors, the incident radiation, with energy greater than or equal to the energy band-gap of the semiconductor, generates majority electrical carriers. The electric conductivity of the material is improved; there is an internal photoelectric effect. Common materials are lead sulfide (PbS), lead selenide (PbSe) and MCT (HgCdTe).

Extrinsic detectors are similar to intrinsic detectors. The difference lies in the fact that carriers are excited from the impurity levels and not over the band-gap of the basic material; this is achieved by doping the semiconductor material. The most-used materials are silicon and germanium doped with impurities such as boron, arsenic and gallium. The spectral response of these detectors can be controlled by the doping level.

Photo-emissive detectors are bi-materials involving a metal layer superimposed to a semiconductor layer; a typical example is platinum silicide (PtSi) on silicon (Si). The process, which is also called external photoelectric effect, consists of the emission of carriers from the metal into the semiconductor under photon absorption. The main advantage is a more uniform response since the response depends on the characteristics of the metal. However, since the photon absorption is proportional to the square of the wavelength, this type of detector is more indicated for long wavelengths.

The QWIP generally consists of layers of thin gallium arsenide alternated to layers of aluminium gallium arsenide (AlGaAs). The basic principle is similar to that of extrinsic detectors with the peculiarity that the dopants are concentrated in microscopic regions and create the quantum wells. The radiation is absorbed by the entire quantum well, not only by a single doping atom, and thus the absorption is increased (so is the response) with respect to extrinsic detectors.

3.4.3 Key Steps in Hardware Evolution

The first prototype was produced at the Swedish Defence Department in cooperation with AGA in the early 1950s; it was based on a thermistor bolometer detector type, which, except for miniaturization, is almost the technology mostly used nowadays. The principle of operation was pretty simple and smart, but the time required to obtain a thermogram was very long, 15 min. Some initial devices, manufactured before World War II, were based on semiconducting diode detectors formed from indium antimonide or lead-tin-telluride materials; the single detector needed mechanical scanning to form a complete image. An airborne optomechanical IR-imaging system developed by Barnes (USA) in 1954 is reported. This opened the way to the development of forward looking IR (FLIR) systems, which were mounted on aircraft and operated with line scanning only, since the other scanning was assured by aeroplane movement.

The system evolved and at the end of the 1950s a second prototype named Thermal Sight (TS1) was presented. To make the image presentation faster, a higher number of sensors was used involving both vertical and horizontal scanning. The demand for higher sensitivity and faster frame rate made the researcher move further on and, at the beginning of 1960s, the Swedish Defence Department (FOA) started to use a new scanning principle together with a fast InSb LN_2 cooled detector [12]. In the meantime, a new scanning concept was designed and patented by FOA and tested by AGA; the novelty resided in a fast-moving refractive horizontal prism. Between 1964 and 1965, AGA presented the very first commercial IR system named AGA Thermovision, mainly developed for predictive maintenance on electrical power lines and after being adapted, using different lens types for other purposes.

Later in the 1970s, a viable alternative to semiconductors was devised in the pyroelectric vidicon tube (produced by Philips and EEV); the target materials were first produced with sensitivities in the shorter wavelength spectral region and afterwards extended to the region between 8 and 14 μm [13,14]. Attempts to make the pyroelectric technology competitive went through improvements regarding the materials [15] and input/output signals [16,17]. In 1978 Raytheon R&D group (part of Texas Instruments) patented ferroelectric IR detectors based on barium strontium titanate [18]. By the end of the 1980s, the attention was on the development of vanadium oxide (VO_x) microbolometer technology, which was disclosed through three patents in 1990–2000 [19–21]. This technology led later to the production of handheld cameras.

From 1970 to 1990, there has been development of IR devices for civil applications by AGA, which later became Agema and then Flir Systems. The Thermovision 870, released by Agema in 1986, represented a milestone since it was the first device equipped with a thermoelectric cooling system and the first of a camera generation that was capable to satisfy the most demanding IR applications for a long period of time. The reason for this success was the full range of detectors, lenses and accessories available. A very important event was the release of the FPA technology by the US military in the late 1980s, which entered the civil market in 1995 with the Thermovision 1000 by Agema. FPA [22] is the technology actually in use coupled with both uncooled and cooled detectors.

The IR devices available today may be classified as belonging to two main families:

1. Uncooled microbolometric detectors, which include a vast variety of handheld cameras from the simplest, cheapest model up to a relatively sophisticated device. They perform well in the maintenance field, which is also called the CM (condition monitoring) field.
2. Cooled detectors that allow for high-sensitivity cameras to be employed in the R&D sector. These are the cameras mostly used for NDE purposes.

3.4.4 Detector Performance

The performance of a detector is evaluated through three parameters: *responsivity* R_ν, *noise equivalent power* (NEP) and *detectivity* D^*.

The responsivity R_v is a measure of the signal output S_{out} (voltage, or current) per incident radiation E_{mc} over the active area of the detector A_{det}:

$$R_v = \frac{S_{out}}{E_{mc} A_{det}} \quad [3.15]$$

The active area is determined as the ratio $A_{det} = A_0/A_e$ between the optical area A_o and the actual *electrical* area A_e.

The NEP defines the intrinsic noise level of the detector, or better the detection limit of the detector. It can be expressed by:

$$\text{NEP} = \frac{N_{out}}{R_v\sqrt{\Delta f}} \qquad [3.16]$$

with N_{out} the noise output and Δf the noise bandwidth.

The detectivity D^* defines the resolving power of the detector and is expressed in terms of the signal-to-noise ratio (S_{out}/N_{out}) with respect to the incident power:

$$D^* = \frac{\sqrt{A_{det}}}{\text{NEP}} \qquad [3.17]$$

The noise that affects detection of IR radiation is of different types and comes from several sources [23]. One major noise source is the electronic circuitry, which gives rise to two different types of noise. One is *thermal noise*, also known as Johnson noise [24], or Johnson—Nyquist noise, which refers to fluctuations in the electrical signal induced by thermal motion of free electrons in a resistive element [25]. Thermal noise is important for thermal detectors while it is generally negligible for photon detectors for which the detection limit is evaluated by accounting for the background radiation. The other type of noise linked to the electronic circuitry is *shot noise*, first described by Schottky in 1918 and often also named Schottky noise. From a general point of view, since electrons share the particle—wave duality with photons, shot noise applies to both. As a definition, shot noise is due to the fluctuations of the number of independent charge carriers crossing a boundary (electrons in an electronic circuit, or photons in an optical device). It is a random process following Poisson statistics (for which the mean squared fluctuation of the number of emission events is equal to the average count N) and the signal-to-noise ratio (SNR) takes the expression:

$$\text{SNR} = \sqrt{N} \qquad [3.18]$$

Of course, shot noise dominates when the finite number of particles that carry energy is small enough. As N increases, the Poisson distribution approaches the Gaussian one and shot noise becomes insignificant. Shot noise is independent of frequency (white noise), but, unlike thermal noise, it is also independent of temperature, meaning it cannot be eliminated by lowering the sensor temperature. Generally, shot noise dominates at high frequency and is linearly proportional to the current [26]. At low frequency (below 10 kHz), another type of noise dominates, which is related to time-dependent fluctuations; this noise is called *flicker noise*, or $1/f$, or pink noise, and, in contrast to the white noise (thermal and shot), is frequency dependent. Flicker noise occurs in virtually all electronic components, as well as in many other physical items in everyday life from the earth's rotation to undersea currents [27,28]. There are two major theoretical hypotheses to its origin. One refers to the fluctuation of the number of carriers caused by capture and emission of carriers into trap centres close to the substrate/dielectric interface. The second hypothesis accounts for fluctuations of the mobility of free carriers, since it is believed to be due to microscopic defects in the semiconductor material.

As described in literature, when the area of the device is small, or the number of defects in the device reduces to one or two, flicker noise becomes *burst noise*. This type of noise also got the name of popcorn noise because the sound it makes when played over a speaker resembles the sound of popcorn popping. Burst noise is described to look, on an oscilloscope, like a square wave of constant magnitude but with random pulse widths. Recently this noise was also referred to as random telegraph signal (RTS). In particular, Kleinpenning [29] showed that RTS noise exists with devices with a small number of carriers, where a single electron can be captured by a single trapping centre. RTS noise is present in submicrometre MOS transistors and in bipolar junction transistors with defected crystal lattice, as well as in modern SiGe transistors. This type of noise is a function of temperature, induced mechanical stress and radiation, and has significant effects at low frequency (typically $f < 1$ kHz). It is described to manifest as a sudden step, or jump, in base current on bipolar transistors, or a step in threshold voltage for FETs. Bursts can happen several times a second or, in some rare cases, may take minutes to occur. They are believed to be caused by charge traps or microscopic defects in the semiconductor material.

In IR detectors, noise comes from two main sources: the IR detector itself with its circuits and the background fluctuations. The first contribution is generally negligible with respect to the second one and then the detection limit is evaluated by accounting only for the background radiation. This contribution is called *background limited IR photodetection* (BLIP). PV detectors are characterized by a $\sqrt{2}$ higher BLIP; in fact, the relationship for a PV detector is:

$$D^*_{\mathrm{BLIPpv}} = \frac{\lambda}{\mathrm{hc}} \frac{\sqrt{\eta_{\mathrm{q}}}}{\sqrt{2\Phi_{\mathrm{b}}}} \qquad [3.19]$$

while for a photoconductive is:

$$D^*_{\mathrm{BLIPpc}} = \frac{\lambda}{2\mathrm{hc}} \frac{\sqrt{\eta_\mathrm{q}}}{\sqrt{\Phi_\mathrm{b}}} \quad [3.20]$$

Φ_b being the background photon flux density (also called dark current). The value of Φ_b received by the detector depends on its responsivity to the wavelengths contained in the radiant source and on its field of view of the background.

A relevant parameter for IR systems is the *noise equivalent temperature difference*; it can be defined as the temperature change, for incident radiation, that gives an output signal equal to the rms noise level N_L:

$$\mathrm{NETD} = N_\mathrm{L} \frac{\Delta T}{\Delta S_\mathrm{m}} \quad [3.21]$$

where ΔS_m is the signal measured for the temperature difference ΔT. It is worth noting that herein only some general information was supplied; for more details the reader should see the related literature.

3.4.5 Quantum Well Infrared Photodetectors

QWIPs deserve particular consideration since some IR cameras designed for the R&D field are based on this technology. QWIP is a semiconductor device generally made of layers of thin GaAs alternated with layers of AlGaAs. GaAs/AlGaAs quantum well devices have a number of potential advantages, including the use of standard manufacturing techniques based on mature GaAs growth and processing technologies; highly uniform and well-controlled molecular beam epitaxy growth on GaAs wafers greater than 6 in.; high yield and thus low cost; more thermal stability and extrinsic radiation hardness [10]. Conversely, QWIPs present some problems: low quantum efficiency ($\eta_\mathrm{q} < 50\%$, typically $\cong 20\%$), lower operating temperature, high dark currents, complex light coupling schemes and narrow spectral response that limits sensitivity in low-background conditions.

As a fabrication drawback, photons normally impinging are not absorbed (do not produce excitation) and so roughed surfaces are needed to scatter photons inside the detector. The major source of noise in QWIPs is the dark current (ie, the current that flows through a biased detector in the dark with no photons impinging on it). As reported by Rogalski [23], at low temperature (<40K) the dark current is mostly caused by defect-related direct tunnelling, while in the medium range between 40 and 70K the thermally assisted tunnelling dominates. At temperatures above 70K, the dark current is mostly dominated by classic thermionic emission (ie, heat-induced flow of charge carriers) of ground-state electrons directly out of the well into the continuum.

The choice of the detector operating temperature is performed as a compromise between reducing the dark current and improving the sensitivity, since the two parameters are correlated. Generally, the choice falls on the value of 70K opting for high sensitivity. From a practical standpoint, QWIPs are suited for large array (640 × 512 pixels) formats, low frame rates and large integration time.

3.5 INFRARED IMAGE GENERATION

The most common detector array types used on commercial IR cameras are microbolometers, for the handheld and automation models, and quantum sensors, for high-performance R&D applications. Along with the detector, another important component of an IR imaging system is the readout circuit. In modern cooled cameras, the detector focal plane assembly is generally tightly attached to the cold side of a Stirling cooler to allow for conductive heat exchange. The Stirling system removes heat from the cold finger and dissipates it at the warm side; it has relatively low efficiency but is adequate for cooling the IR camera detector.

A typical FPA can be described as a two-dimensional matrix of n columns by m rows of individual cells, or *pixels*, with each pixel (detector) having micrometre-size dimensions. The interface circuit (or driving circuit) of each cell, which converts the current from each photodiode to a voltage at each pixel, is referred to as a *readout integrated circuit* (ROIC) unit cell. A control circuit is also needed to synchronize the operation of the unit cells and converts the output of the IR pixels to a stream of bits [30]. The main function of an ROIC is transforming the small diode incremental current, generated by IR radiation, into a relatively large measurable output voltage; this is commonly done by integrating the photocurrent in a capacitor during a fixed period of time. There are two integration modes: the *asynchronous integrate while read* (AIWR) and the *asynchronous integrate then read* (AITR). In the AIWR mode, the camera is able to integrate scene signals while simultaneously reading out the previous frame. This mode allows for semi-independent control of the camera's frame rate and integration time. The frame rate is the number of snapshots taken per second; the integration time is the length of time that each snapshot actually views the scene. However, the time required to read out all of the pixels limits the maximum frame rate; the selection of a smaller window size allows the maximum frame rate to increase. The AITR mode is used in some radiometric cameras coupled with specific calibrated windows [31,32].

Two main signals that are used for time synchronization of the circuit are line synchronization (LSYNC) and frame synchronization (FSYNC), which along with the master clock (CLK), multiplex the provided array of the unit cell voltages to a serial stream, sequentially. The master clock is the highest frequency clock and provides a reference for pixel readout; the other clocks (FSYNC and LSYNC) are integral numbers of CLK periods. In particular, the LSYNC controls the readout synchronization of each individual line of the array, while the FSYNC marks the beginning of the frame time. Starting from the first cell of the first row, the outputs of all cells in the row go to the column multiplexer. With each clock pulse, the multiplexer connects the output of one of the cells in the first row to the output stage. After multiplexing the last cell in the row, the LSYNC pulse triggers the row shift register and the second line is selected (the column shift register is also reset). Similar to the first row, each column will again be selected sequentially by the clock pulse. At the end of the frame, the FSYNC pulse resets both row and column shift registers and the pointer returns to its initial position to start a new frame. The LSYNC and FSYNC pulses are generated based on the ROIC operation mode (integrate-while-read or integrate-then-read). More specifically, in any case, the rising edge of the FSYNC triggers the beginning of the frame, but it is followed by a sequence of LSYNC signals for every line change in the AIWR mode making the frame time approximately equal to the read time. Conversely, in the AITR, the FSYNC signal remains high until the end of the readout sequence, and the integration starts just after the readout, resulting in a frame time that is approximately equal to the readout time plus the integration time [33].

3.5.1 Some Problems Affecting Infrared Images

Every pixel in the sensor array is acting like a single element detector with its own thermal response curve (offset and gain). This means that if we put in front of the camera an object with a homogeneous temperature distribution, we would like to see each pixel of the array giving out the same signal, in terms of digital level, or grey level or colour level. Instead, a raw image may appear affected by nonuniformity and may be almost unusable without the due software corrections. Common effects are:

- *Fixed pattern noise* is due to spatial intensity values observed either as a gridlike pattern or as a striping pattern laid on the top of the true image. It is due to the pixel reading procedure in the IR cameras that implement FPA.
- *Bad pixels*, defined as anomalous pixels that can be distinguished in dead pixels, permanently dark, and hot pixels, permanently white. A map of dead pixels is generally known from the FPA manufacturer.
- *Vignetting*, which appears as a darkening of the image corners while the centre remains brighter. It is related to the gap between the temperature of the scene and the temperature of the lens.
- *Different offset and gain of the pixels*, which is almost unavoidable in the detector production process.

All these effects deteriorate the image quality, hence, to get a good image out of a FPA detector, it is necessary to go through the characterization of each array pixels. This process is the so-called "*Nonuniformity Correction*" mapping. This is part of the calibration procedure and in the microbolometer camera is factory prepared and stored in the camera memory. For laboratory systems instead, this process can be either stored in the camera memory or performed by the user according to the working conditions. The procedure consists of placing in front of the camera two homogeneous black bodies, one at low temperature and another one at higher temperature. All the image pixels are scanned by the software and checked one by one for offset, gain and wrong response. Gain and offset values are identified and then stored in file maps corresponding to the image size. The nonresponding pixels are instead stored in another file map; the bad pixels will be replaced by the neighbours. Once calculated, these maps will be applied on the fly to the raw image at camera scanning speed.

Another problem may be the camera's ambient temperature drift, or better the radiation caused by the heating and cooling of the camera itself, mainly of the internal electronic circuitry. Any swings in camera internal temperature caused by changes in environment, or the heating and cooling of camera electronics, will affect the radiation intensity at the detector. This is called *parasitic radiation* and can cause inaccuracies in the camera output, especially with thermographically calibrated cameras. Certain IR cameras have internal sensors that monitor changes in camera temperature and send feedback to the camera processor. The camera output is then corrected for any parasitic radiation effects; this functionality is commonly referred to as *ambient drift compensation* [34].

3.5.2 Measurement Accuracy

The measurement accuracy depends on many factors involving amongst others the instrument performance and the thermographer's skill. Of course, the magnitude of temperature differences to be measured is of great concern.

In fact, some limitations may arise when attempting to measure temperature variations that are very small, in particular, at the edge of the instrument resolution. In this regard, a difficult task is to perceive the temperature variations that are related to thermoelastic effects, eg, developing under relatively low loads. This is because of the presence of the detector noise; practically, the amplitude of the signal to be measured may fall within the noise amplitude. Some detectors may exhibit random jumping noise that masks the temperature variations to be measured; this may happen when the temperature variations are linked to thermoelastic effects. It has been recently demonstrated by the authors [35] that it is possible to account for and eliminate most of the detector noise in a simple and effective way with the use of a reference area.

As already mentioned, knowledge of the thermal emissivity coefficient ε is very important for an accurate temperature measurement with an IR-imaging system. In fact, deficient knowledge of ε not only affects the radiation emitted by the measured surface but also that which, coming from the environment, is reflected by said surface ($\rho = 1 - \varepsilon$). As well, painting the object's surface to increase its emittance cannot be implemented when transient heat transfer is involved due to the thermal resistance of the painting itself, which is not generally known. However, as it will be later shown, this concern is not important with phase images obtained by means of lock-in thermography.

DISCLOSURE

Some figures, which refer to a well known theory, were adapted from Ref. [11].

SUMMARY TO CHAPTER 3 AND INTRODUCTION TO CHAPTER 4

The basic relationships have been described through an historical excursus from discovery to detection and deployment of IR radiation. In particular, the IR devices actually in use, detector types and working implications were discussed. Some general information about the detector performance was also supplied. Now, the work proceeds with practical applications of IRT as nondestructive evaluation technique in Chapter 4 and as online monitoring device of loading events to follow in Chapter 5.

REFERENCES

[1] J.C. Maxwell, A dynamical theory of the electromagnetic field, Philos. Trans. R. Soc. London (1865) 459–512.
[2] G.S. Smith, An Introduction to Classical Electromagnetic Radiation, Cambridge University Press, 1997.
[3] V. Singh, Einstein and the quantum special section: the legacy of Albert Einstein, Curr. Sci. 89 (12) (2005) 2101–2112.
[4] ASTM E1933-97, Standard Test Methods for Measuring and Compensating for Emissivity Using Infrared Imaging Radiometers, American Society for Testing and Materials, Pennsylvania, USA, 1997.
[5] H. Levinstein, J. Mudar, Infrared detectors in remote sensing, Proc. IEEE 63 (1975) 6–14.
[6] S. Soloman, Sensors Handbook, McGraw Hill Professionals, 1998.
[7] P. Norton, J. Campbell, S. Horn, D. Reago, Third generation infrared imagers, Proc. SPIE 4130 (2000) 226–236.
[8] M. Henini, M. Razeghi (Eds.), Handbook of Infra-red Detection Technologies, Elsevier, 2002.
[9] A. Rogalski, P. Martyniuk, InAs/GaInSb superlattices as a promising material system for third generation infrared detectors, Infrared Phys. Technol. 48 (2006) 39–52.
[10] A. Rogalski, Competitive technologies of third generation infrared photon detectors, Opto-Electron. Rev. 14 (2006) 87–101.
[11] C. Meola, Origin and theory of infrared thermography, Chap 1, in: C. Meola (Ed.), Infrared Thermography: Recent Advances and Future Trends, Bentham Science Publishers, 2012, pp. 3–28. eISBN 978-1-60805-143-4.
[12] G.T. Nichols, Temperature Measuring, US2008793, 1935.
[13] W.S. Beller, Navy sees promise in infrared thermography for solid case checking, Missiles Rochets 16 (1965) 22.
[14] D.R. Green, Principles and applications of emittance-independent infrared nondestructive testing, Appl. Opt. 7 (1968) 1779–1786.
[15] ASTM E2582-07, Standard Practice for Infrared Flash Thermography of Composite Panels and Repair Patches Used in Aerospace Applications, 2007.
[16] H.S. Carslaw, J.C. Jaeger, Conduction of Heat in Solids, Clarendon Press, UK, 1959.
[17] R. Rinaldi, Chapter 2: Infrared Devices: Short History and New Trends, in: C. Meola (Ed.), Infrared Thermography: Recent Advances and Future Trends, Bentham Science Publishers Ltd., 2012.
[18] P.W. Kruse, L.D. McGluaghlin, R.B. Mcquistan, Elements of Infrared Technology, Wiley & Sons, New York, 1962.
[19] R.D. Hudson, Infrared Systems Engineering, Wiley-Interscience, New York, 1969.
[20] L.E. Garn, E.J. Sharp, Pyroelectric vidicon target materials, IEEE Trans. Parts Hybrids Packag. 10 (1974) 208–221.
[21] J.T. Hall, Electronic Motion Compensation for the Pyroelectric Vidicon, US4040087, 1977.
[22] V.T. Bly, Optical System for Thermal Imager, US4051370, 1977.

[23] A. Rogalski, Infrared detectors: status and trends, Prog. Quantum Electron. 27 (2003) 59–210.
[24] J.B. Johnson, Thermal agitation of electricity in conductors, Am. Phys. Soc. 32 (1928) 97.
[25] E.L. Dereniak, D.G. Crowe, Optical Radiation Detectors, John Wiley & Sons, 1984, pp. 39–40.
[26] C. Beenakker, C. Schönenberge, Quantum shot noise, Phys. Today 56 (2003) 37–42.
[27] P. Dutta, P.M. Horn, Low-frequency fluctuations in solids: 1/f noise, Rev. Mod. Phys. 53 (1981) 497–516.
[28] A.A. Balandin, Noise and Fluctuations Control in Electronic Devices, American Scientific Publishers, Los Angeles, 2002.
[29] T.G.M. Kleinpenning, On 1/f noise and random telegraph noise in very small electronic devices, Physica B 164 (1990) 331–334.
[30] M. Singh, M. Sadana, S. Sachdev, G. Pratap, Development of miniature Stirling cryocooler technology for infrared focal plane array, Def. Sci. J. 63 (6) (2013) 571–580.
[31] H.S. Jatana, D. Sehgal, S. Sachan, K. Dhiman, R.K. Tripathi, A. Yadav, Readout Integrated Circuit (Roic) for 8 × 8 QWIPS Array. Technical Report, Semiconductor Complex Ltd., Sector 72, S.A.S. Nagar, Chandigarh, Punjab, India, 2006.
[32] Flir Systems SC6000 Manual.
[33] N.Y. Aziz, R.F. Cannata, G.T. Kincaid, R.J. Hansen, J.L. Heath, W.J. Parrish, S.M. Petronio, J.T. Woolaway, Standardized high performance 640 by 512 readout integrated circuit for infrared applications, in: Proc. SPIE 3698, Infrared Technology and Applications XXV, vol. 766, July 26, 1999, http://dx.doi.org/10.1117/12.354577.
[34] http://www.flir.com/uploadedFiles/Thermography/MMC/Brochures/T559243/T559243_APAC.pdf, (accessed on August 2014).
[35] C. Meola, S. Boccardi, G.M. Carlomagno, Measurements of very small temperature variations with LWIR QWIP infrared camera, Infrared Phys. Technol. 72 (2015) 195–203.

CHAPTER 4

Nondestructive Testing With Infrared Thermography

4.1 HISTORICAL GROUNDINGS

Infrared thermography (IRT) has taken its first steps right in the nondestructive testing (NDT) field. In fact, signs of its use for NDT tasks were traced by Vavilov [1] up to the beginning of the last century when the first infrared detector was patented by Parker in 1914 [2]. The proposal for using a complete IR system for detection of forest fires came from Barker in 1934 [3]. In the meantime, Nichols performed one of the first industrial applications [4], which consisted of checking the heating uniformity of steel strips. An innovative study on thermal nondestructive testing (TNDT) was performed by Green, who developed a technique for checking nuclear reactor fuel elements [5]. In the 1960s, the very first commercial infrared system (AGA Thermovision) entered the market; it was first used for the inspection of electric/electronic components and later implemented for nondestructive purposes such as testing of the Polaris rocket motor [6] and nuclear reactor fuel elements [7]. In the meantime, IRT also started to become attractive for use in the aerospace field.

In 1966 the American Society for Nondestructive Testing (ASNT) published the first edition (ASNT SNT-TC-1A) of the Recommended Practice to provide guidelines for personnel involved in nondestructive tests. In this first edition, the document involved only five methods. ultrasonic testing, magnetic testing, liquid penetrant testing, eddy current testing and radiographic testing. IRT was added much later, in 1992, as an emergent technique and it was fully recognized in 2007 when the ASTM E2582 practice, concerning application of flash thermography for inspection of aerospace composite panels, was released [8].

However, the first approaches were somewhat deceiving, being rather qualitative, so interest was turned toward other NDE techniques such as ultrasonics. This happened mainly because of lack of basic knowledge regarding the output of the infrared imaging device and the thermal behaviour of the part under inspection. At that time there were no well-assessed testing

Infrared Thermography in the Evaluation of Aerospace Composite Materials
ISBN 978-1-78242-171-9
http://dx.doi.org/10.1016/B978-1-78242-171-9.00004-8

procedures and the interpretation of thermograms was rather difficult. It was only later (beginning in the 1980s), thanks to the comprehension of heat transfer mechanisms [9], that IRT received renewed attention. A thermophysical approach has been developed by many scientists (amongst others, Carlomagno, Balageas, and Vavilov) who introduced one- (1D), two- (2D) and three-dimensional (3D) models of defects detection. From that time on, there has been a proliferation of ideas regarding both hardware and software means for setting up an effective technique, which is now amongst the well-recognized NDE techniques in the aeronautical field. For more details, the reader is addressed to international standards, like the NAS 410, or ISO 9712.

4.2 SOME BASIC CONCEPTS

When using the term nondestructive inspection, with the general meaning of discovering a fault in a generic component without destroying it and without compromising its normal operation, we may be facing two types of undertakings:

- ascertaining the proper functioning of a device when it goes to operation
- assessing the absence of defects inside the component material

Also depending on the type of addressed task, two basic approaches can be used: active or passive.

4.2.1 Passive Thermography

Passive thermography is carried out on components that in operating conditions become either hot or cold as a prerequisite of their proper functioning. The inspection does not require any external thermal stimulation of the object. In other words, the heat flow, due to the intrinsic heat generation of the object under test, is exploited for assessing a correct functioning, or a malfunctioning, of that object. For example, passive thermography may be used to inspect furnaces, electrical appliances, switchboards or refrigerators. Due to Ohm's law, an increase of electric resistance induces an increase of temperature. Then, it is possible to distinguish overheated connections amongst many, as well as defects associated with other problems (Fig. 4.1).

Indeed, an infrared imaging device can be fruitfully exploited for predictive maintenance to discover a problem before it becomes crucial and,

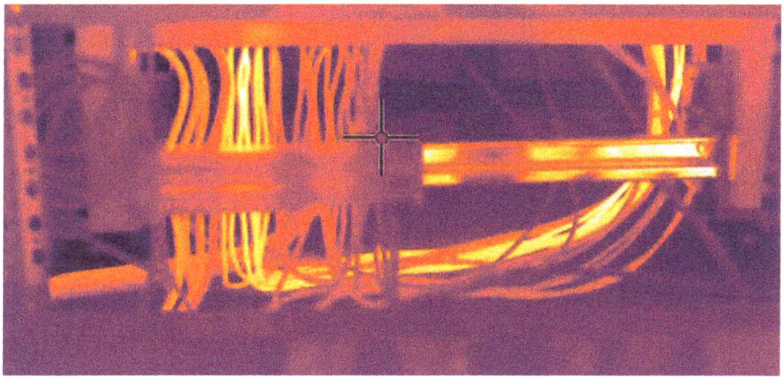

Figure 4.1 Electrical connections inside a switchboard (see the overheated wires).

often, expensive. In fact, before a component fails, generally it heats up because of energy dissipation; excess heat is usually the first sign of trouble in electrical or mechanical apparatuses. Such troubles may be due to loose connections, overloaded or imbalanced circuits, defective breakers, damaged switchgears, faulty fuses, material defects, excessive friction and to many other conditions, which may lead to severe overheating and component failure. An infrared imaging device is the best diagnostic tool available to quickly identify and fix potential problems with breakers, interconnections, wire crimps, damaged bearings, worn-out gears and loose or corroded connections; this mostly occurs before the problem causes safety hazards and productivity losses. Owing to the fact that electrical distribution systems are often the foundation for many others, a failure in an electrical system may cause communication, fire safety, security and mechanical systems to fail as well.

An infrared imaging system is also advantageously used for electrical power line surveys; this task can be also addressed by using either helicopters or aeroplanes. Nowadays, even drones are employed, eg, for the inspection of photovoltaic cell systems.

In the aerospace field, passive thermography can be suitable for assessing the presence of water inside fuselage panels of landing aircraft.

Apart from predictive maintenance, passive thermography may be advantageously exploited for finding hot spots in thermoelastic stress analysis and locating damaged areas. This task will be discussed in detail in Chapter 5.

It is worth noting that often people refer erroneously to passive thermography when dealing with inspection of building envelopes. They make this error because they think they are not using any means for thermal stimulation while, in fact, they are often using the sun, or in any case the weather changes. Indeed, the sun is exploited for heating up a surface and for performing active thermography testing!

4.2.2 Active Thermography

In *active thermography*, the inspection regards an object, which is initially in thermal equilibrium with the surroundings, and then a certain amount of energy is introduced in it to produce thermal contrast to highlight the features of interest. Nondestructive inspection of materials with an infrared imaging system generally consists of thermally stimulating the object (under exam) and monitoring of its surface temperature variation during the transient heating, or cooling, phase. The analysis is generally performed by relating the stimulation to the heat conduction within the object [9].

The relativistic heat conduction theory predicts that the most general equation governing the temperature variation under thermal stimulation of a material is Cattaneo's equation [10]:

$$Q + t_r \frac{\partial Q}{\partial t} = -k\nabla T \tag{4.1}$$

with Q the heat flux, T the temperature, t the time and t_r and k the material relaxation time and the thermal conductivity coefficient, respectively. By recognizing in it the conservation of energy (first thermodynamics law), Eq. [4.1] can be rearranged to have the hyperbolic (telegraph) equation:

$$\frac{1}{V_h^2}\frac{\partial^2 T}{\partial t^2} + \frac{1}{\alpha}\frac{\partial T}{\partial t} - \nabla^2 T = 0 \tag{4.2}$$

with α and V_h, respectively, the material thermal diffusivity coefficient and speed of propagation of thermal waves:

$$V_h = \sqrt{\frac{\alpha}{t_r}} \tag{4.3}$$

When $V_h \rightarrow \infty$, or $t_r \rightarrow 0$, ie, in the case of zero relaxation time (infinite heat propagation velocity), abandoning the relativistic theory, the hyperbolic Eq. [4.2] becomes the classical parabolic *Fourier's heat conduction equation*:

$$\frac{\partial T}{\partial t} - \alpha\nabla^2 T = 0 \tag{4.4}$$

Therefore, the heat propagation through a material is generally driven by its static thermal properties.

4.2.3 Thermal Properties of Materials

Materials are characterized by different thermal properties and behave in a different manner when heated. Following is a list of the most relevant useful thermal properties of a material.

Thermal conductivity: It is an intensive property (bulk property) and is generally designated with the symbol k. It indicates the ability of a material to conduct heat and is equal to the ratio of the heat flux $\dot{q}$ (energy per unit area and unit time) divided by a temperature gradient ∇T (temperature difference per unit length $\Delta T/L$):

$$k = \frac{L\dot{q}}{\Delta T} \qquad [4.5]$$

The thermal conductivity depends on the material molecular configuration and on its thermodynamic state. The most general definition foresees a thermal conductivity that is a tensorial quantity. For instance, pure crystalline substances exhibit very different thermal conductivities along different crystal axes, due to differences in phonon coupling along a given axis. Sapphire, or more simply wood, is a notable example of thermal conductivity variable with the heat flux direction (nonisotropic). In the absence of convection phenomena, air and other gases are generally isotropic and good insulators (low k). Therefore, many insulating materials are simply obtained by superimposing a large number of small gas-filled pockets that prevent large-scale convection. The reciprocal of thermal conductivity is *thermal resistivity*. Often, the thermal conductance is used, which may represent the quantity of heat that passes per unit time through a given plate when the plate's opposite faces differ in temperature by one degree. The reciprocal of thermal conductance is *thermal resistance*.

Thermal diffusivity: It indicates the rapidity with which a material adjusts its temperature to that of its surroundings. It is generally indicated with the symbol α and is defined as the ratio between the thermal conductivity k and the product of density ρ (mass per unit volume) times specific (per unit mass) heat capacity c (specific heat):

$$\alpha = \frac{k}{\rho c} \qquad [4.6]$$

Specific heat is the measure of the energy Q necessary to raise a unit mass m of material by one degree of temperature ΔT:

$$c = \frac{Q}{m\,\Delta T} \qquad [4.7]$$

The product ρc is also named *specific heat capacity per unit volume* and represents the thermal property that is present in Fourier's equation (Eq. [4.4]).

Thermal effusivity: It is generally indicated with the symbol e and is a measure of the material ability to exchange thermal energy with its surroundings:

$$e = \sqrt{k\rho c} \qquad [4.8]$$

Sometimes, the thermal effusivity is also called *thermal inertia*.

Some clarifying examples

Let us consider a slab (Fig. 4.2), which is initially ($t = 0$) at uniform ambient temperature (T_a), that is to say $T_1 = T_2 = T_a$, where T_1 and T_2 are the temperatures of the front slab side 1 and of the back side 2, respectively. If the slab is exposed on side 1 to a uniform heat source for a certain time period, two cases may be considered, depending on the slab thermophysical characteristics:

Case I: The slab is made of a homogeneous material without any defect. After some time, the temperature on side 2 starts to increase uniformly over this surface, until a steady state T_s is asymptotically reached (Fig. 4.3). The final temperature T_s depends on the total heat input and the slab thermal capacity; both quantities can also be considered per unit surface. Instead, the time t_s, at which the temperature on the unheated side gets to a quasi-constant value, depends on the slab thickness and material

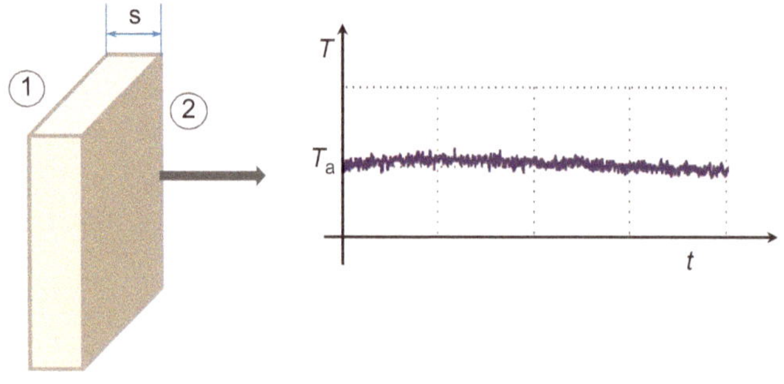

Figure 4.2 Plate at ambient temperature.

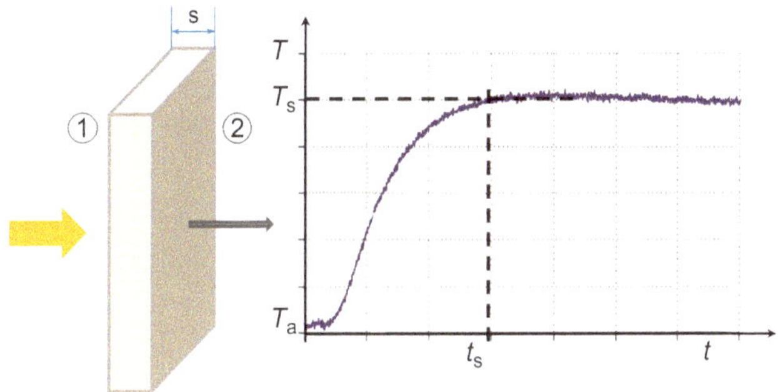

Figure 4.3 Temperature rise on side 2 when heating up of the plate from side 1, for a certain time.

thermal diffusivity α. Being s the slab thickness, the time t_s can be roughly estimated as $t_s = s^2/\alpha$, so it is directly proportional to the square of the slab thickness.

Case II: A defect is buried inside the material, as shown in Fig. 4.4. Again, two cases may arise depending on the value of the thermal diffusivity of the defect α_d with respect to that of the bulk sound material α_b.

1. $\alpha_d < \alpha_b$ – because of its lower thermal diffusivity, the defect acts as an obstacle to the heat propagation. Then, the temperature distribution on side 2, which is initially uniform, will show, for certain time, a minimum in correspondence of the defect (Fig. 4.4(a)). Sometimes after t_s, at thermal equilibrium, the slab temperature will become again thoroughly uniform.
2. $\alpha_d > \alpha_b$ – heat propagation is increased through the defected zone. Then, the temperature distribution on side 2, which is initially uniform, will show, for a certain time, a maximum in correspondence of the defect (Fig. 4.4(b)). Sometimes after t_s, at thermal equilibrium, the slab temperature again will become thoroughly uniform.

During the temperature evolution, defects may be identified as local temperature differences (contrasts) in the surface temperature map:

$$\Delta T = T_s - T_d \qquad [4.9]$$

with T_s the temperature in the sound material zone and T_d the temperature over the defect. There is a time instant for which the absolute value of ΔT attains its maximum, ie, defects are visualized with the best contrast. It is worth noting that defects can be visualized only during a transient phase; going toward steady-state conditions, $t \gg t_s$, $\Delta T \rightarrow 0$ and it is no longer possible to detect defects.

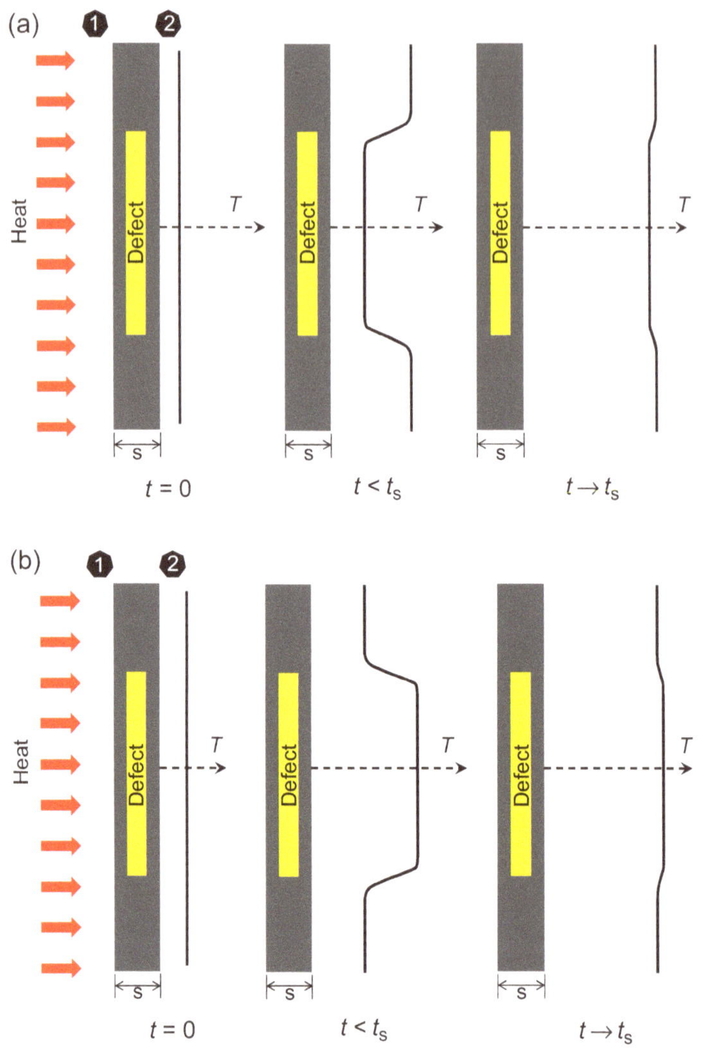

Figure 4.4 Temperature profiles evolution on side 2 for (a) $\alpha_d < \alpha_b$ and (b) $\alpha_d > \alpha_b$.

4.3 THERMOGRAPHIC NONDESTRUCTIVE TESTING TECHNIQUES

As a general premise, the use of IRT for NDE basically requires the sound material having relatively uniform and known thermal characteristics, such as conductivity, specific heat and diffusivity. In fact, in the presence of a thermal irregularity, defects, like disbonds, delaminations and slag inclusions, can be detected from their anomalous thermal signature with respect to the sound base material.

Basically, two thermographic techniques can be used for nondestructive inspection: *pulse thermography* (PT) and *lock-in thermography* (LT). There are some other techniques, such as pulse phase thermography (PPT), time-resolved-infrared (TRIR), vibrothermography etc., which practically are modifications of PT and LT, because they include either a different heating method and/or a different processing algorithm.

4.3.1 Pulse Thermography

PT simply consists of stimulating the object under evaluation with a heating pulse and monitoring its surface temperature variation during the transient heating or cooling phase. Heating generally should be performed with flash lamps, but it can also be accomplished with cheaper halogen lamps. In the last case, heating requires a longer time and the technique is named *long-pulse thermography* (LPT) or *step pulse thermography* (SPT). Often, the term PT is also used to indicate LPT, without any distinction. Thermal stimulation can be performed likewise with other means like scanning lasers or hot air jets. Thermal stimulation can also be achieved by cooling instead of heating, and this is mainly realized by means of cold air jets. Of course, air jets (hot or cold) can be used only on relatively massive objects since jet impingement may impair delicate surfaces, and their use does not generally produce a uniform heat flux over the surface.

The slab analysis with PT can be performed in two different modes (Fig. 4.5): *transmission* (Fig. 4.5(a)) and *reflection* (Fig. 4.5(b)). In the transmission mode, the infrared camera views the object rear face, ie, the face

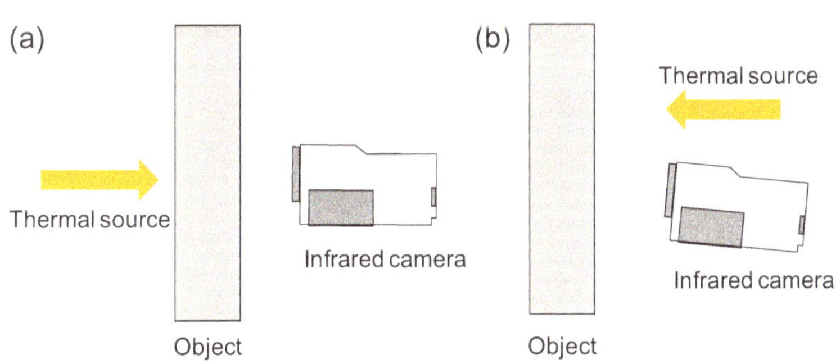

Figure 4.5 Set-ups for pulse thermography (long-pulse thermography) tests. (a) In transmission mode (b) In reflection mode.

opposite to the heated/cooled side. Since this face is not always accessible and/or available (eg, the external surface of an aircraft), the reflection mode, for which both heating/cooling source and camera are positioned on the same side, is frequently applied.

The thermal energy propagates by conduction inside the material while the infrared camera monitors the temperature variation over the viewed surface. Obviously, for a uniform surface heating and homogeneous material, the temperature distribution is uniform (Fig. 4.6(a)). The presence of a defect at a certain depth interferes with the heat flow causing local surface temperature variations (Fig. 4.6(b) and (c)). In particular, thermal images shown in Fig. 4.6 refer to glass epoxy specimens, of which the left one is made of sound material while the other two include three Teflon inserts that, simulating defects, appear as dark stains in the temperature map. This is possible to notice as the shallower inserts (Fig. 4.6(b)) are better outlined than the deeper ones (Fig. 4.6(c)). Thermal images of Fig. 4.6 were taken during heating up with halogen lamps, in transmission mode, with the Agema Thermovision 900 LW.

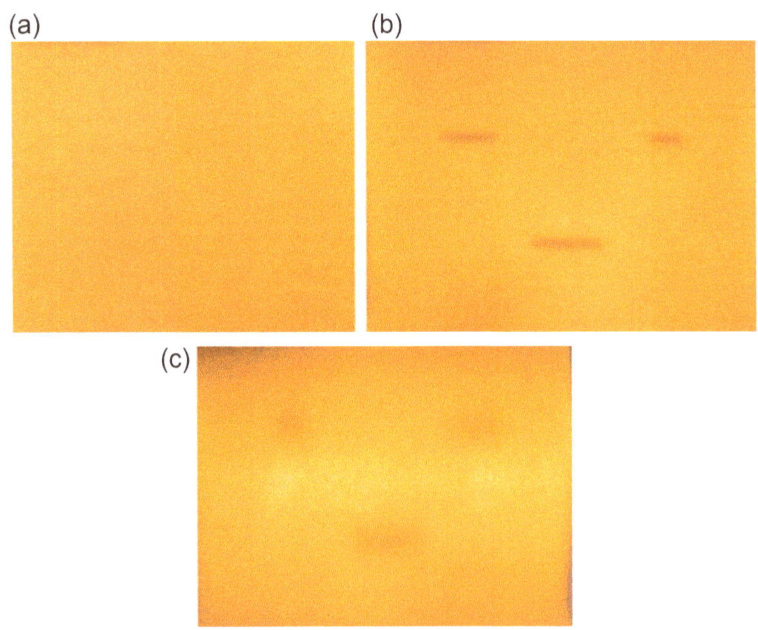

Figure 4.6 Thermal images of sound and defective materials. (a) Sound material (b) Buried defect at $p = 1$ mm (c) Buried defect at $p = 4$ mm.

By referring to Fig. 4.6(c), profiles along the two upper defects are plotted in Fig. 4.7 (line A) for several times (starting from $t = 0$). According to the plots, each stain first appears with a milder under temperature (which is true especially for deep defects) and, after, is better outlined at a certain time instant. For the tested conditions and as can be seen from Fig. 4.7, owing to defects located at $p = 4$ mm, the best contrast is attained after about 30 s. In particular, Fig. 4.7 shows the temperature rise with respect to the initial ambient value ($\Delta T = T - T_r$). More specifically, T_r (reference temperature) is the temperature of the sound material before starting of transient heating $T_r = T_{(t=0)}$ (ie, the temperature of the first thermal image taken at $t = 0$ s in the time sequence).

The visibility of defects can be evaluated by the parameter (*dimensionless contrast*) D_T [11]:

$$D_T = \frac{|\Delta T_d|}{|\Delta T_s|} = \frac{|T_s - T_d|}{|T_s - T_r|} \qquad [4.10]$$

where: T_d is the temperature over the defective zone, T_s is the temperature in the sound zone and T_r is the reference temperature. Indeed, the quantity $T_s - T_r = \Delta T_s$ plays an important role because it indicates the optimal temperature variation to which the material has to be subjected in order to have good defect detection. In Eq. [4.10] the dimensionless temperature difference D_T is evaluated owing to absolute ΔT_d and ΔT_s values since they can be either positive or negative, their sign depending on the test procedure (acquisition during heating or during cooling) and on the thermal

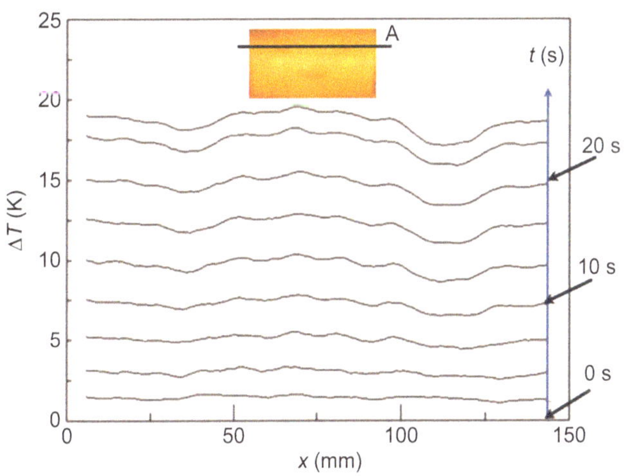

Figure 4.7 Thermal profiles of sound and defective materials ($p = 4$ mm).

properties of both the sound material and the material of which the defect is made. Generally, a deep defect becomes visible later than a shallow one and a larger defect produces a more marked temperature difference. Fig. 4.8 shows a plot of the maximum dimensionless contrast D_T against the defect depth p for glass/epoxy specimens with Teflon inserts.

Thus, a complete (from surface to depth) nondestructive evaluation of a material is performed by acquiring thermal images (thermograms) in a time sequence. The thermal image may be shown in grey or colour, levels on the computer monitor (10, 16,..., 112, or more), each level corresponding to a temperature interval; such interval depends on the camera selected range (the minimum allowable range for the scanner is often about 2 degrees). Information about the defect parameters such as size, depth and thermal resistance can be obtained by applying postprocessing procedures to the thermograms.

4.3.1.1 Thermal Contrast and Defect Detectability

The visibility of defects depends on several factors, which include material characteristics (ie, thermal contrast), environmental conditions and instrument sensitivity. As regards the material aspect, both thermal properties (thermal conductivity, thermal diffusivity) and defect geometry play key roles [12].

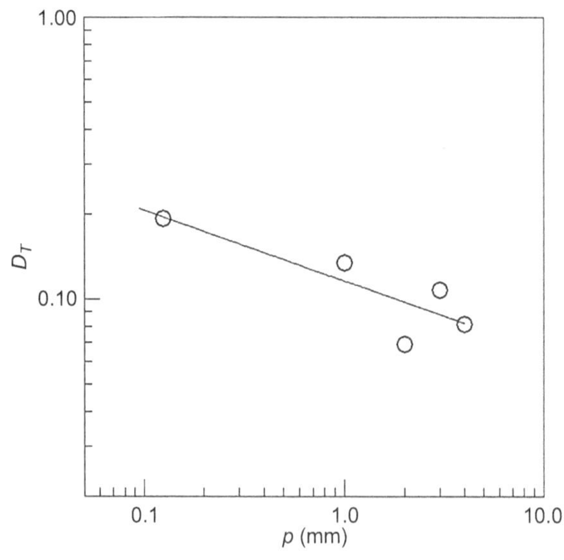

Figure 4.8 Maximum contrast as a function of defect depth.

By introducing the dimensionless relative thermal diffusivity α_r defined as the difference between the thermal diffusivity of the sound material α_s and that of the defect α_d with respect to α_s

$$\alpha_r = \alpha_s - \frac{\alpha_d}{\alpha_s} \quad [4.11]$$

it may be affirmed that the thermal contrast increases by increasing the absolute value of α_r. Of course, as α_r decreases, meaning that the thermal properties of the defect fall within those of the sound material, the thermal contrast drops ($\Delta T \rightarrow 0$ in Eq. [4.10]) and the defect detectability becomes crucial. However, the case of a very thin delamination in a composite material, or what it is referred to with the more general term of *kissing bond*, cannot be fully described by α_r.

Of great relevance is, of course, the depth at which a defect is located. It is generally assumed that the loss of thermal contrast is proportional to the cube of the defect depth. In particular for homogeneous isotropic materials, Maldague [13] proposed an empirical rule of thumb for which the radius of the smallest detectable defect should be at least one to two times larger than its depth under the surface. By analyzing composite materials with artificially inserted defects, Meola et al. [14] found that the thickness of the defect is also very important. In particular, a large radius defect could be hardly distinguishable if it is very thin (such as kissing bonds); conversely, a defect of smaller radius, but of larger thickness, may be more easily detectable.

As far as the instrument sensitivity is concerned, the defect detection is limited by the camera signal-to-noise ratio [15]:

$$\mathrm{SNR} = \frac{\overline{T}_d - \overline{T}_s}{\sigma_s} \quad [4.12]$$

where $\overline{T}_d$ and $\overline{T}_s$ are the average temperature values in the defective and sound areas, respectively, and σ_s is the standard temperature deviation in the sound area. It is obvious that a defect to be detectable must entail a ΔT signal greater than the noise level. The noise is due in part to the detector, in part to the inspected material and in part to boundary conditions, such as the heating parameters (exposure time, intensity, distribution over the surface) as well as to a lesser extent to the environmental conditions. In addition, the defect features that appear on the surface during pulsed thermography tests are always affected by three-dimensional heat conduction within the material, resulting in misleading determination of size and depth of defects.

4.3.1.2 Estimation of Defect Size

In PT measurements, the surface finish is of great importance since variations in surface roughness, cleanliness, uniformity of existing paint and other surface conditions can cause variations of the emissivity (and absorptivity) coefficient, thus affecting the temperature measurement. These aspects were analysed by Meola and Carlomagno [11]. The main problem, which is encountered in the measurement of the size of a defect, especially from thermal images obtained with LPT, is the difficulty to discriminate the interface between sound and defective zones. More specifically, the temperature difference ΔT (in absolute value) over a damaged zone, with respect to the sound material, generally attains its maximum value over the centre of the defect and decreases laterally going toward the sound material in a kind of bell-shaped way, as sketched in Fig. 4.9. This effect increases in the presence of overheated and underheated areas on the surface. To avoid these features, attention has to be paid to the choice of the type of excitation source, the surface orientation as well as the distance between the specimen and the stimulation source.

A criterion to discriminate between sound material and defect may be to establish the minimum acceptable ($\Delta T = T_s - T_d$) value by considering both the heat transfer mechanisms and the noise signal [12,15]. By assuming T_m as the average temperature over sound material, and T_c as the value on the surface above the centre of the defect, it is possible to define the defect contour as the locus of local values T that satisfy both of the following conditions [11]:

$$\begin{cases} \dfrac{T_m - T}{T_m - T_c} \cong 0.5 \\ T_m - T > \Delta T_\sigma \end{cases} \qquad [4.13]$$

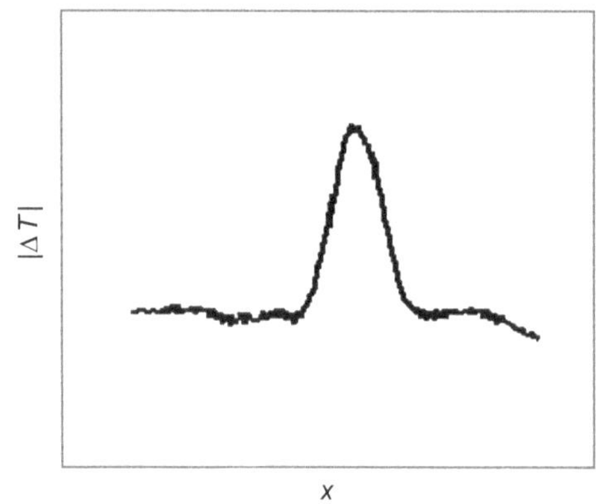

Figure 4.9 $|\Delta T|$ distribution over a defective area.

ΔT_σ being the average temperature deviation over the sound material. This also means that a defect is detectable and measurable only if its associated signal exceeds the noise level.

4.3.1.3 Estimation of Defect Depth

The defect depth can be estimated owing to deviations from the ideal evolution of surface temperature corresponding to the energy input [16]. For a semi-infinite, sound sample, the time-dependent surface temperature response to an instantaneous heat pulse is:

$$T(t) - T(0) = \frac{Q}{k\rho c\sqrt{\pi t}} \qquad [4.14]$$

with Q the energy input per unit area. Eq. [4.14] applies to one-dimensional heat conduction, instantaneous thermal flash and absence of heat losses within the sample. When the heat applied to the surface, propagating into the sample, encounters a subsurface discontinuity, a pronounced deviation of the temperature difference dependence from linearity with $1/\sqrt{t}$, described by Eq. [4.14], is found (Fig. 4.10). The time required for this deviation from the ideal behaviour to occur is a function of the depth of the interface through the simple rule:

$$t \propto \frac{z^2}{\alpha} \qquad [4.15]$$

which relates the thermal propagation time t to the defect depth z and the material thermal diffusivity α. Starting from Eq. [4.14], many methods were developed within the past 15 years including registered patents [17–20].

The patent US5711603 by Ringermacher et al. [18] disclosed a method to determine the defect depth z from the contrast peak. Later, in 2003, the patent US6542849 by Sun [19] proposed a curve-fitting method for prediction of the defect depth, which is based on a theoretical heat transfer

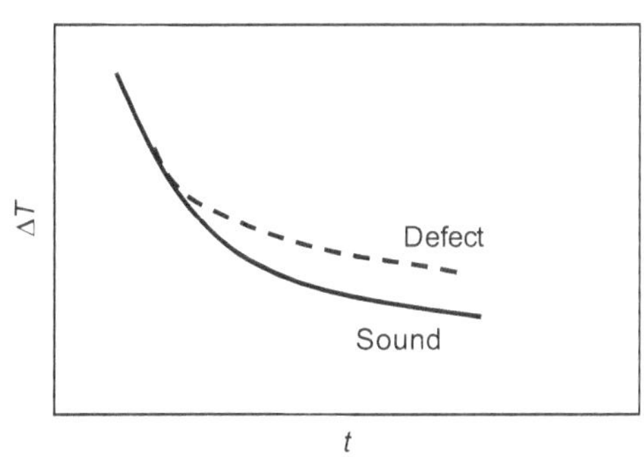

Figure 4.10 Surface time–temperature decay.

model to fit the temporal thermography data at each surface point. In 2006, some methods, such as the peak temperature-contrast and peak temperature-slope, the logarithmic peak second-derivative and the least-square fitting methods, were analysed by Sun [20] for the case of a ceramic composite flat plate including several machined flat-bottom holes to simulate defects.

4.3.1.4 Estimation of Material Thermal Diffusivity

Under one-dimensional heat propagation assumption and once the sample thickness is known, Eq. [4.15] can also be used to determine the slab material thermal diffusivity. This is the so-called *flash method*, which was proposed by Parker in 1961 [21] and became an ASTM standard in 1992 [22]. More specifically, this method is based on the analytical solution of Fourier equation for an infinite surface slab, subjected to a thermal Dirac pulse, with adiabatic boundary conditions at its back side [9]. The test is performed according to the transmission scheme (Fig. 4.5(a)). In particular, an energy pulse is irradiated on the sample front surface, while the temperature evolution is measured on the back surface (Fig. 4.3). Thermal diffusivity is linked to the sample thickness L and to the half-maximum temperature rise time $t_{1/2}$ through the equation:

$$\alpha = \frac{0.139L^2}{t_{1/2}} \qquad [4.16]$$

where $t_{1/2}$ represents the time needed to reach half of the final maximum temperature value. Of course, initially this method was conceived in conjunction with contact temperature transducers (eg, leaf thermocouples), while now it can be also used in combination with IRT temperature detection. The usefulness of the infrared camera is obvious since it allows, in a remote way, for fast monitoring of temperature variations and for evaluation of the thermal diffusivity, pixel by pixel, over a given, ever large area. The possibility to map the thermal diffusivity is advantageous in view of accounting for its local variations due to the presence of local material inhomogeneities. Once the material thermal diffusivity is known, flash thermography can be alternatively exploited to measure the slab thickness (patent EP1173724) [23].

4.3.1.5 Evaluation of Material Porosity

It is known that the presence of porosity in composites may affect their mechanical properties, and the high probability to induce formation of voids within composites during their manufacturing processes. Hence, there is the need for availability of NDT techniques able to discover the presence and the distribution of porosity in final parts. In recent years, flash thermography has emerged as one of the most valuable inspection methods and is currently used for the evaluation of the apparent thermal diffusivity (Eq. [4.16]). Bearing in mind that the formation of voids produces a small decrease of the apparent density, but especially the decrease of thermal conductivity because of the lack of material and, being the thermal diffusivity linked to these two properties (Eq. [4.6]), the evaluation of the thermal diffusivity may lead to the estimation of the amount of embedded porosity within the inspected material. In fact, both properties decrease, but the influence of the thermal conductivity drop prevails on that of the density.

To this end, flash thermography is now an almost mature technique after 15 years of investigation. The main developed steps were reviewed by Meola and Toscano [24]. Herein, only some of the most recent works are recalled. In particular, Hendorfer et al. [25] experimentally derived a linear relation between diffusivity and porosity. They performed measurements on carbon fibre-reinforced polymer (CFRP) samples 2 mm thick with simulated porosity between 1% and 5%, the average nominal porosity being estimated through ultrasonic testing (UT). Flash thermography has proved to have some advantages over ultrasonics in terms of detailed information supplied and reduced testing time involved. Mayr and Hendorfer [26] investigated also the possibility to apply flash thermography in the reflection mode and found a higher data spread with respect to using the transmission mode. In particular, they observed a decrease of sensitivity for both modes (transmission and reflection) for porosity increasing up to 10%; this was ascribed to the tendency of pores to interlink, as assessed through measurements with computed tomography (CT). Later, Mayr et al. [27] carried out tests on CFRP, involving prepregs woven in a twill weave pattern at a ratio of 3/1, and found active thermography comparable to UT in terms of ability to determine porosity. As a main finding, they observed the thermal diffusivity to be strongly affected not only by the porosity content but also by the pores' shape, which was established through comparisons with CT.

The usefulness of flash thermography within the evaluation of porosity in composites seems extensively demonstrated. However, most of the studies deal with qualitative estimation of porosity; some refer to composites of given characteristics not always completely specified, which make a general data correlation difficult. On the other hand, before a full acceptance of the method, many tests are still needed, as well as comparison of data coming from different laboratories worldwide and carried out on similar test articles. This latter point is very difficult to achieve, owing to the variability associated with composites. In fact, composites are made of two basic ingredients: the matrix and the fibres, but each of them may have different thermal properties; besides, it is their mutual percentage that drives the thermal diffusivity of the final part [27]. For example in carbon/epoxy, carbon fibres have thermal diffusivity much higher than epoxy resin; hence, the thermal diffusivity of a CFRP part strongly depends on the percentage and orientation of carbon fibres. Through flash thermography measurements, Wrobel et al. [28] found a variation of thermal diffusivity in the range of $1.5-2.2 \times 10^{-7}$ m^2/s for a variation of fibre content between 13% and 28%. As already mentioned, fibres orientation also plays a fundamental role since the easiest heat propagation direction coincides with the direction of the fibres [29]. Finally, the manufacturing process involves so many parameters that it is difficult, and almost impossible, to replicate a product with completely identical characteristics.

A quite extensive investigation was carried out by Toscano et al. [30,31] involving a large number of CFRP samples of several different stacking sequences and with different percentages of porosity. In particular, by plotting the apparent thermal diffusivity against porosity (Fig. 4.11) it is possible to see first a rapid decrease of α for porosity increasing from 0% up to 3% and after a much milder reduction following a power law as porosity overpasses 5%. At first look, this would mean that flash thermography loses its ability to discriminate small increases of porosity in porous composites. However, this cannot be a problem in advanced composite for aerospace applications, where the void content is quite critical and levels above about 1% cannot be tolerated.

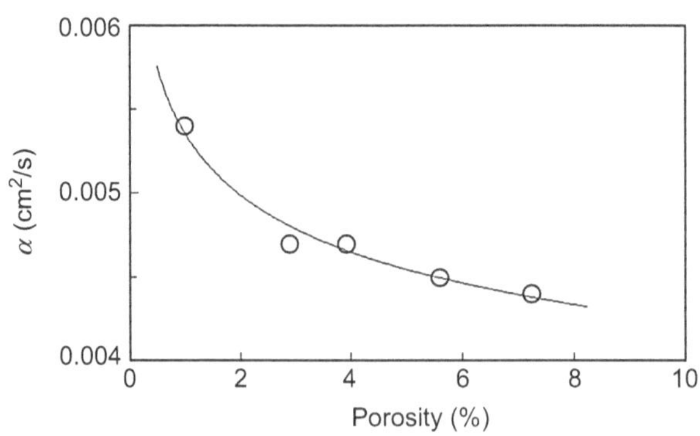

Figure 4.11 Apparent thermal diffusivity against porosity.

Results obtained with flash thermography were also compared with results obtained with UT [24]. It has been demonstrated that the measure of thermal diffusivity by flash thermography, as a technique for porosity evaluation, may be an alternative to the typical ultrasonic attenuation estimation. However, flash thermography offers some advantages being very effective, noncontact (no coupling media are necessary), fast and not affected by the surface finish; this means that, unlike UT, a part can be inspected by indifferently viewing its smooth or rough side. In addition, flash thermography allows to contemporaneously detect manufacturing defects and assess the porosity amount within only one test, with consequent economic advantages. The test set-up can be easily implemented in the industrial processes for in-line inspection of parts. Finally, it is important to note that NDT with IRT is carried out using simple and safe (for the personnel) set-up arrangements, with further advantages in terms of safety-at-work concerns.

4.3.2 Lock-in Thermography

The concept of lock-in thermography was first introduced by Carlomagno and Berardi in 1976 [32] and later further investigated by many other researchers [33–36]. Classical lock-in thermography is basically performed with a modulated heating lamp and is more properly called *optical lock-in thermography* (LT or OLT; some authors also call it *modulated thermography*, MT). The energy, generally irradiated by halogen lamps, is delivered to the object surface in the form of a periodic thermal wave. The thermographic system is coherently coupled to the thermal wave source, which is operated in such a way that a sinusoidal temperature modulation of the object surface results. This modulation is obtained from a nonlinear electrical signal produced by the so-called lock-in module, which allows also for frequency variation. The system collects a series of thermal images and compares the modulated heating to the measured temperatures by extracting the sinusoidal wave pattern at each point of the image.

By considering photothermal methods applied to the lock-in amplifier (which allows filtering the stationary component), the harmonic heat delivered at the surface ($z = 0$) of a homogeneous and semi-infinite material results in a (time-dependent) thermal wave, which propagates inside the material according to the relationship:

$$\widetilde{T}(z,t) = \frac{\eta Q_i}{2e\sqrt{\omega}}\exp\left(-\frac{z}{\mu}\right)\exp i\left(\omega t - \frac{z}{\mu}\right) = A(z)\exp i(\omega t - \phi(z)) \qquad [4.17]$$

where $\widetilde{T}(z,t)$ is the periodic temperature component, η represents the absorbed part of the incident heat Q_i, e is the material thermal effusivity, $A(z)$ is the wave amplitude, $\phi(z)$ is the phase shift of the thermal wave travelling inside the material and μ is the thermal diffusion length, which is calculated from the thermal diffusivity α and the wave frequency $f = \omega/2\pi$, as:

$$\mu = \sqrt{\frac{\alpha}{\pi f}} \qquad [4.18]$$

In the classical optical lock-in method, both lamp and camera must be located on the same side. The thermal wave propagates inside the material and gets reflected when it reaches parts where the heat propagation parameters change (inhomogeneities). The reflected wave interferes with the surface wave, producing an oscillating interference pattern that can be measured in terms of amplitude $A(z)$ or phase angle $\phi(z)$, which, respectively, produce amplitude or phase images. Amplitude A and phase ϕ at each pixel can be evaluated from four thermographic images ($S1$ to $S4$) taken during one modulation cycle (Fig. 4.12) [36]:

$$\begin{aligned} A(z) &= \sqrt{(S_3 - S_1)^2 + (S_3 - S_1)^2} \\ \phi(z) &= \tan^{-1}\left[\frac{S_3 - S_1}{S_4 - S_2}\right] \end{aligned} \qquad [4.19]$$

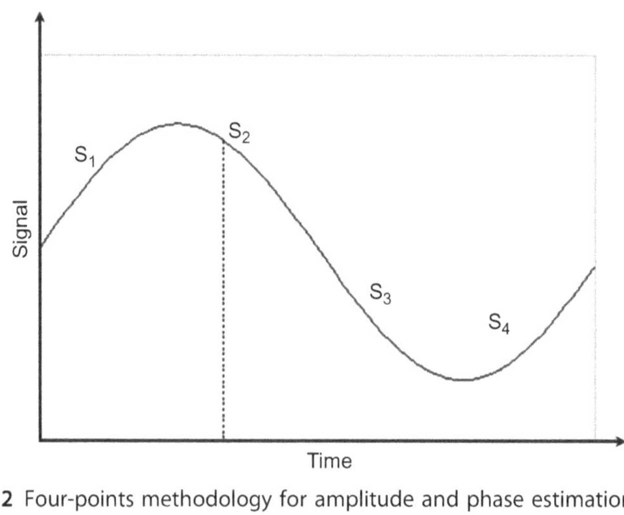

Figure 4.12 Four-points methodology for amplitude and phase estimation by lock-in thermography.

The amplitude is insensitive to superposition of unmodulated radiation caused by reflection effects; the phase angle is also independent of local power density and thermal emissivity coefficient; this means that LT shows some advantages over PT. However, it is worth noting that the four-points method (Eq. [4.19]) applies only for purely sinusoidal stimulation conditions.

The depth range for the amplitude image is given by μ, while the maximum depth p, which can be reached for the phase image, corresponds to 1.8μ [33–35]. Therefore, the material thickness, which can be inspected starting from the stimulated surface, depends on the wave period (the longer the period, ie, the lower the frequency, the deeper the penetration) and on the thermal material properties (thermal conductivity and volumetric specific heat). The variation of the depth p with f is shown in Fig. 4.13 for four different materials, specifically: carbon/epoxy (CFRP), glass/epoxy glass fibre-reinforced polymer (GFRP), Plexiglas and polyvinyl chloride (PVC). It is possible to notice, as higher depth values are reached, that CFRP has the highest thermal diffusivity. In addition, values for the other three materials (GFRP, PVC and Plexiglas) that have close α values practically lie over the same correlation curve.

Generally, tests start at a quite high wave frequency, at which only surface (or low-depth) defects are visible and, later on, to inspect deeper layers, the frequency is decreased until the entire specimen thickness has been traversed or the minimum selectable frequency has been reached.

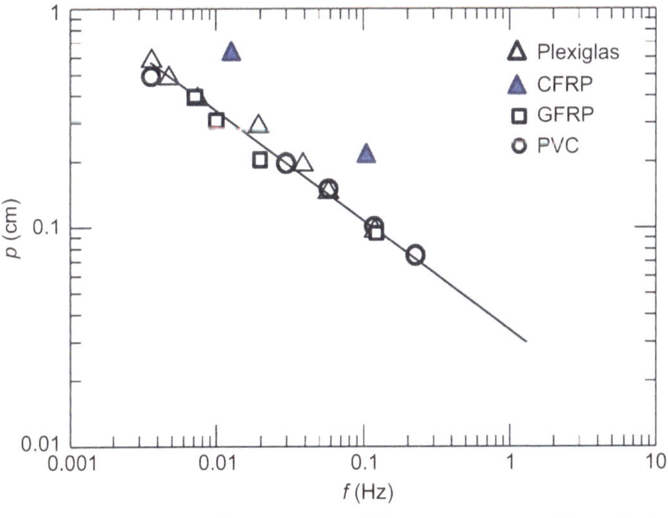

Figure 4.13 Depth against frequency for different materials. CFRP; GFRP; PVC.

Results are generally presented as phase images where a local variation of colour indicates a local variation of phase angle ϕ and, in turn, a local variation of material thermal properties.

4.3.2.1 Estimation of Defect Size

The size of a defect can be evaluated owing to local variations of surface temperature and using the previous established criterion [11] to discriminate between sound and damaged materials in flash thermography, as described by Eq. [4.13], which can be properly rewritten as:

$$\frac{\phi_m - \phi}{\phi_m - \phi_c} \cong 0.5 \quad \text{for } \phi_m - \phi > \Delta\phi_s \tag{4.20}$$

where ϕ is the phase angle in a generic point, ϕ_m is the average phase value for sound material, ϕ_c is the value above the centre of the discontinuity and $\Delta\phi_s$ is the average deviation of ϕ over the sound material. The conversion from pixels to physical units is easily done once the spatial resolution of the employed optics is known.

4.3.2.2 Estimation of Defect Depth and Thickness

As described in patents US4878116 and DE4203272 [37,38], LT offers some advantages over PT, in terms of noise reduction, for the evaluation of the wall thickness. In general, LT allows for simple and fast defects characterization without much troublesome postprocessing procedures. In fact, the phase angle can be assumed as a characteristic material parameter [39] and Eq. [4.18] supplies information about defects' depth and thickness [11]. The only requirement is the knowledge of the thermal diffusivity α in Eq. [4.18], but the latter can be evaluated by LT itself, following the procedure described in Meola et al. [39]. More specifically, for a given material under exam, one has to consider a homogeneous specimen (without defects) of known thickness s_p and to find the frequency f_p for which the entire thickness s_p is traversed. By introducing the two values f_p and s_p in Eq. [4.18], which is rearranged in the form:

$$\alpha = \pi f_p \left(\frac{s_p}{1.8}\right)^2 \tag{4.21}$$

with $s_p/1.8 = \mu$, the searched α value can be obtained. This represents a fast way to obtain the overall average thermal diffusivity of composites that is not always available since, as already mentioned, it is influenced by several factors such as the type of material (fibre and matrix), the volume of fibres with respect to that of the matrix and the fibres orientation and stacking sequence. Eq. [4.20] can be rearranged in terms of phase angle contrast:

$$C_\phi = \frac{\phi_m - \phi}{\phi_m - \phi_c} \quad [4.22]$$

Bear in mind that $\Delta\phi$ ($\Delta\phi = \phi_m - \phi$), or ($\phi_m - \phi_c$) may change sign with varying the wave frequency (ie, $\Delta\phi$ may become positive, zero or negative). More specifically, when the thermal wave (of suitable frequency value) touches the defect top surface, the defect appears, but with poor contrast (only a mark) and $\Delta\phi$ could be either positive or negative. Generally, when decreasing the frequency the contrast enhances and becomes the highest as the thermal wave approaches the defect centre across its thickness. Then, it worsens again, vanishing ($\Delta\phi = 0$) when the thermal wave reaches the defect bottom, the corresponding frequency being called *blind frequency* f_b. After that, on the phase image, the defect appears again but with a change of colour (change of $\Delta\phi$ sign). These variations of the phase angle may be exploited for the evaluation of the defect thickness. A typical variation of $\Delta\phi$, with an initial negative value, with varying the thermal wave frequency, is represented in Fig. 4.14. Of course, the blind frequency depends on the material thermal properties.

An example of blind frequency is shown in the phase images of Fig. 4.15 for a GFRP specimen with enclosed defects located at $p = 1$ mm; these images were taken with the Agema Thermovision 900 LW. As can be seen, no defects are visible for $f = 0.5$ Hz; defects appear as dark stains for $f = 0.25$ Hz, then they disappear for $f = 0.05$ Hz and appear again as clear stains for $f = 0.025$ Hz. The heating frequency of $f = 0.05$ Hz is clearly the blind frequency looked for.

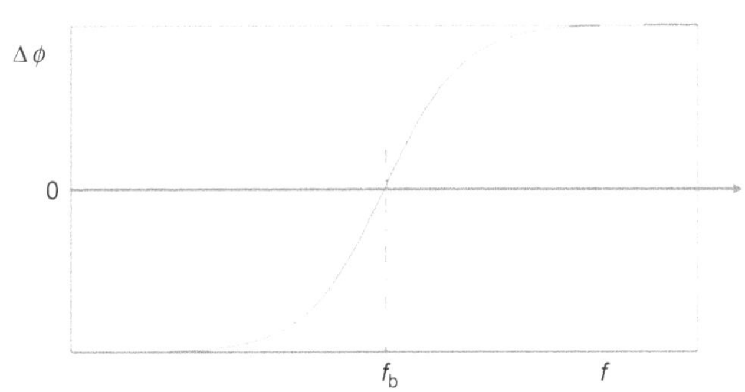

Figure 4.14 $\Delta\phi$ distribution with f.

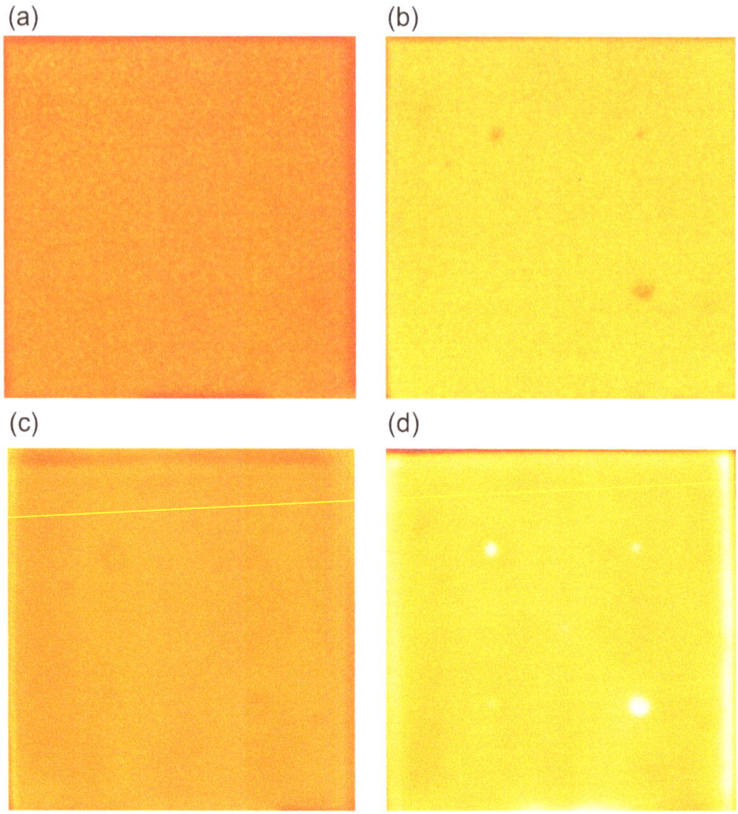

Figure 4.15 Phase images of a GFRP specimen for varying *f*. (a) $f = 0.5$ Hz; (b) $f = 0.25$ Hz; (c) $f = 0.05$ Hz; (d) $f = 0.025$ Hz.

4.3.2.3 LT Testing in Unsteady-State Conditions

Nondestructive evaluation with lock-in thermography should be generally performed in steady-state conditions, but in some circumstances this situation may be unfeasible. Thus, in this section it is described as an alternative method based on the transient approach.

Considering a specimen at room temperature, the application of an oscillating thermal excitation leads, in the first stage, to a kind of staircase-like temperature increase (Fig. 4.16). Indeed, the sinusoidal excitation caused by the alternating switching on/off of the lamps causes a recurrent heating/nonheating condition, respectively. The heat conduction within the material and, above all, the heat exchange with the environment determine the final average temperature at which the specimen after a

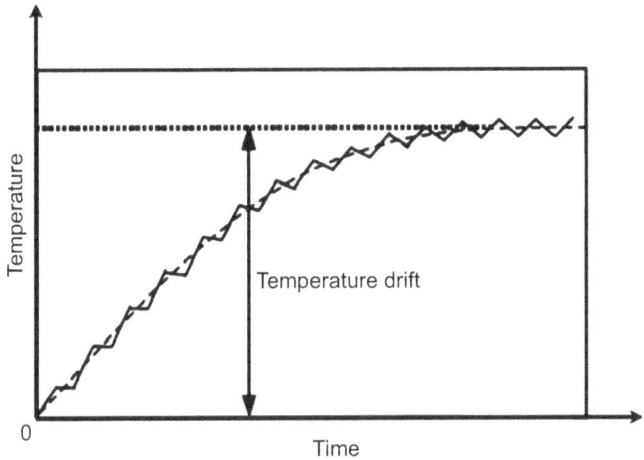

Figure 4.16 Typical temperature evolution during lock-in thermography measurements.

certain time, called *thermal relaxation time*, stabilizes. At this stage, the amount of heat introduced during each period equals the heat loss in the whole period. From then on, a quasi-steady-state condition is reached and the periodic temperature oscillation is superimposed to the reached average temperature.

The thermal relaxation time depends on the heating power and frequency, the specimen thermal capacity as well as the environmental testing conditions; it can range from a few seconds up to many minutes. By definition, a *thermally infinitely thick specimen* will never reach steady-state conditions, although such a circumstance does not practically occur. Indeed, since in LT only the oscillating part of the temperature signal is of interest, a quasi-equilibrium state should be reached. However, this would require a long measurement time and a large data storage capacity. Hence, it seems convenient to record the data even during the unsteady-state conditions and then apply a correction method.

In the initial heating phase, the digitized temperature signal within each lock-in period may be considered as given by the sum of two terms [40]:

- the direct current (d.c.) component, ie, the average value during the period, which causes an almost linear average slope component
- the alternate current (a.c.) component, ie, the oscillating part

Afterwards, the slope of the average temperature profile decreases until, after the thermal relaxation time, it becomes null.

Mathematically, the lock-in signal treatment, also called *lock-in correlation procedure*, consists of successively multiplying the incoming IR images by a set of weighting factors and summing up the results in a frame storage. The weighting factors may approximate a harmonic function and be synchronized with the pulsed bias applied to the sample [40]. This weighted sum can be performed during the measurement and is also referred to as *standard lock-in correlation*.

There are other correlation methods such as the *least squares method* [41] and the *discrete Fourier transform* (DFT). Both of these methods require the availability of all data prior to calculation and so they have to be performed off-line. The DFT includes real and imaginary parts of the complex transform, which are used to estimate the amplitude and the phase [42]:

$$A_n = \sqrt{\mathrm{Re}_n^2 + \mathrm{Im}_n^2}$$
$$\phi_n = \tan^{-1}\left(\frac{\mathrm{Im}_n}{\mathrm{Re}_n}\right) \qquad [4.23]$$

The DFT can be used with any waveform; it succeeds to discern the a.c. component from the d.c. one and to evaluate only the basic harmonic of the signal, but is not able to distinguish the slope component. For these reasons, the amplitude and the phase images obtained from the sequence acquired during unsteady-state condition can be defined as uncorrected.

The simplest approach to obtain a pure oscillating signal is to subtract a polynomial function, which approximates the temperature ramp, from the signal recorded in unsteady regime. Unfortunately, this method involves every frame of the sequence and is inappropriate in the presence of a high number of acquired periods.

Breitenstein et al. [43] proposed a faster correction method, which was initially conceived for electronic device testing. Within this method, the errors related to the temperature drift do not depend on the slope of the ramp but only on the temperature difference between the first thermogram (at the time t_0) and the last one (at time t_{fin}). This assumption holds for a linear ramp as well as for any arbitrarily shaped one, as long as it is composed of a number of linear fractions. As a consequence, the operation time is drastically reduced and is independent of the number of total frames. As an example, Fig. 4.17 displays the raw apparent temperature signal in red with the superimposed third-order polynomial in blue.

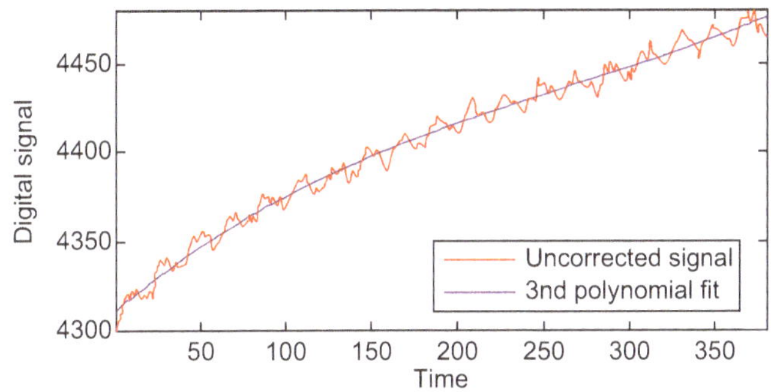

Figure 4.17 Raw signal and pixel-wise third-order polynomial fit for $f = 0.5$ Hz and $N = 19$.

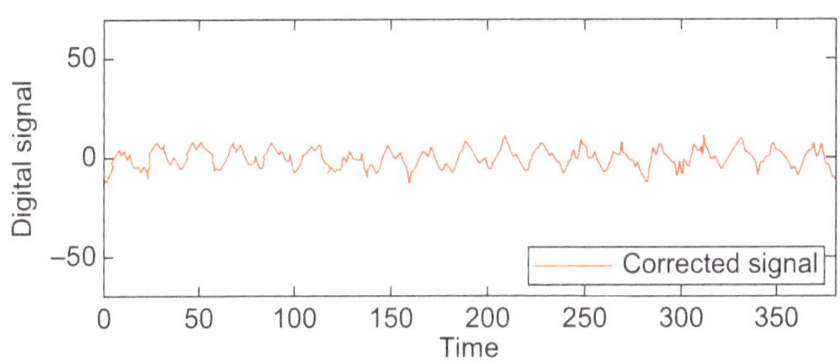

Figure 4.18 Corrected signal after the subtraction of the time domain temperature trend.

After subtraction of the third-order polynomial from the raw signal, the corrected signal in Fig. 4.18 appears as if it had been recorded under steady-state conditions. Furthermore, the d.c. component is automatically suppressed.

4.4 EXAMPLES OF MATERIALS INSPECTION WITH LOCK-IN THERMOGRAPHY

By applying lock-in thermography it is possible to detect most of the defects that may occur in composites as produced by manufacturing processes, as well as damages intervening in service.

4.4.1 Detection of Manufacturing Defects

Amongst defects occurring during manufacturing, LT is able to detect many of them. Some examples are reported in the phase images shown in the following Figs 4.19–4.25. A major concern for all manufacturers is to prevent contamination of the material from spurious particles, but despite efforts in cleaning the workplace, some residuals may still exist. Often fragments of the backing film remain entrapped, which may act as initiator sites once the component becomes a load-bearing structure; such fragments are generally very thin and difficult to be detected.

The phase image shown in Fig. 4.19(a) refers to a CFRP part with an embedded thin plastic piece; apart from the clearly recognizable inclusion, it is also possible to distinguish fibres orientation and consistency. A common problem, especially in laminates fabricated with the hand lay-up technology, is the fibres misalignment as displayed by the phase image shown in Fig. 4.19(b). Note that the hand lay-up method involves a compression force, often applied with a special roller tool, in order to remove as much as possible the air contained between the layers. Such a compression operation may cause two effects: fibres misalignment (Fig. 4.19(b)) and a packthreads effect (Fig. 4.20), as well nonuniformities in the matrix mainly due to the materials soft state.

As shown by the phase images of Figs 4.21–4.23, taken for varying the heating frequency, nonuniform distribution of either fibres or matrix, is commonly encountered when dealing with composite panels. In particular, in the phase images of Fig. 4.22(d), the presence of blistering can also be observed. The most critical occurrence is certainly the formation of porosity, which is mostly driven by the curing process. In fact, curing involves combined effects of temperature and pressure. Temperature is needed to activate and to control the chemical reactions in the matrix

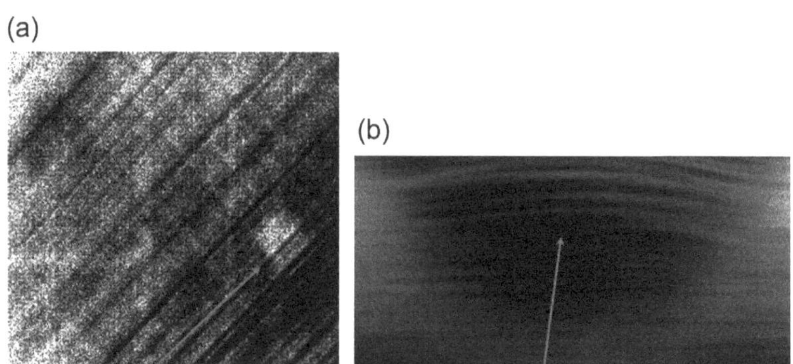

Figure 4.19 Slag inclusion and fibres misalignment in CFRP. (a) Slag inclusion (b) Fibres misalignment.

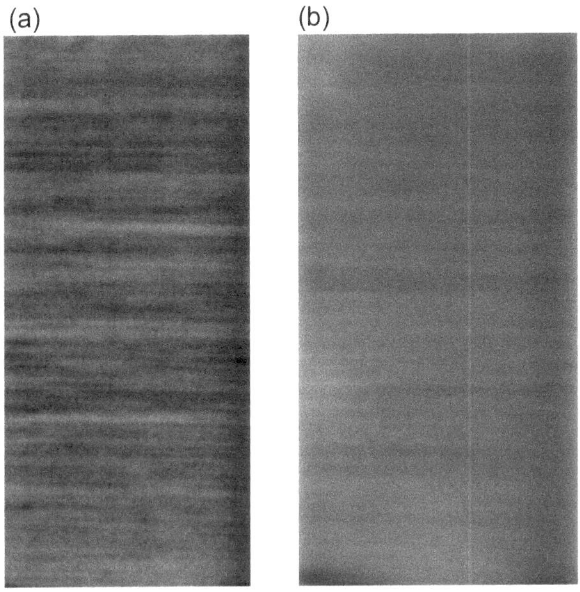

Figure 4.20 Packthreads effect in CFRP. (a) $f = 0.3$ Hz; (b) $f = 0.1$ Hz.

(resin), while pressure is used to squeeze off the excess resin, to consolidate the stacked plies and to minimize the amount of entrapped gas between the plies and within the resin [44]. So, the action of temperature and pressure, as well as the duration of the cycle, are the main parameters to care for in order to assure the effectiveness of the curing process and, as a consequence, the overall quality of the final component.

Indeed, if the pressure level and duration are not optimized, voids could rise in the laminate causing reduction of the interlaminar shear strength [45]. While the presence of a certain amount of voids cannot be completely avoided, it is important to know their acceptable level. In fact, in many composite applications, such as in advanced composite dynamic aerospace structures, the void content is quite critical and levels above about 1% are not tolerable; however, in other applications levels of 5% and higher still can be accepted [45]. Of course, lower voids content means higher production costs, whereas loosening the quality control standards translates to a lower cost of end products. Often, establishing the acceptable level of voids is a critical issue in designing composite structures, and then the use of effective evaluation techniques is compulsory.

(a) (b) (c) (d)

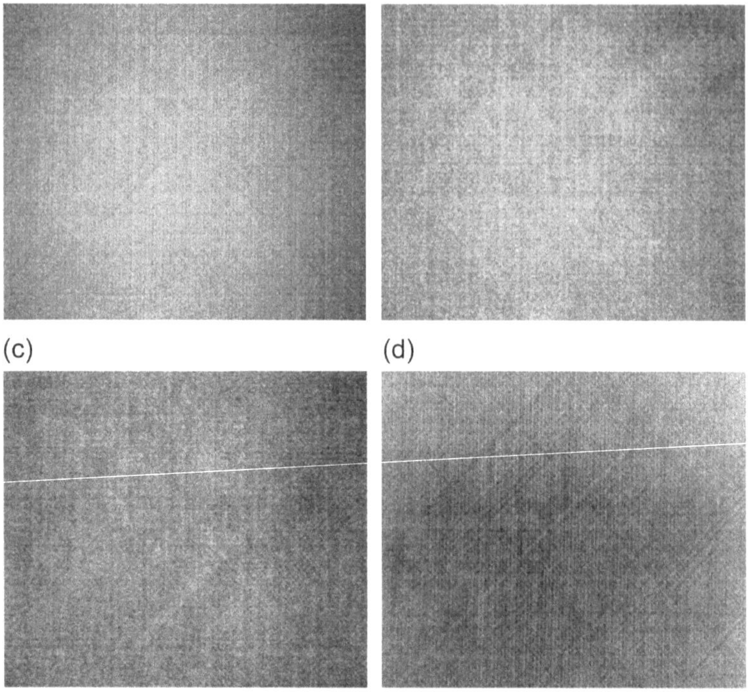

Figure 4.21 Phase images of specimen CFRP-A. (a) $f = 0.87$ Hz; (b) $f = 0.65$ Hz; (c) $f = 0.47$ Hz; (d) $f = 0.30$ Hz.

The phase images shown in Figs 4.24 and 4.25, taken in both reflection (Fig. 4.24) and transmission (Fig. 4.25) modes, supply examples of contemporaneous presence of porosity (darker spots in Fig. 4.24 and lighter in Fig. 4.25) and fibres misalignments. These two defects have detrimental effects on the material resistance to impact, as experimentally demonstrated [46] with the help of IRT.

It is worth noting that all the defects like slag inclusions, delamination, voids, porosity, matrix and fibres nonuniformity entail local variation of material thermal diffusivity and so they can also be discovered with flash thermography [24].

4.4.2 Degradation of Materials in Service

Every part undergoes deterioration in service due to a combination of factors, which mainly include fatigue and adverse environmental conditions. The fatigue life of a structure strongly depends on the type of material it is made of. The failure, under loading, of a metallic material

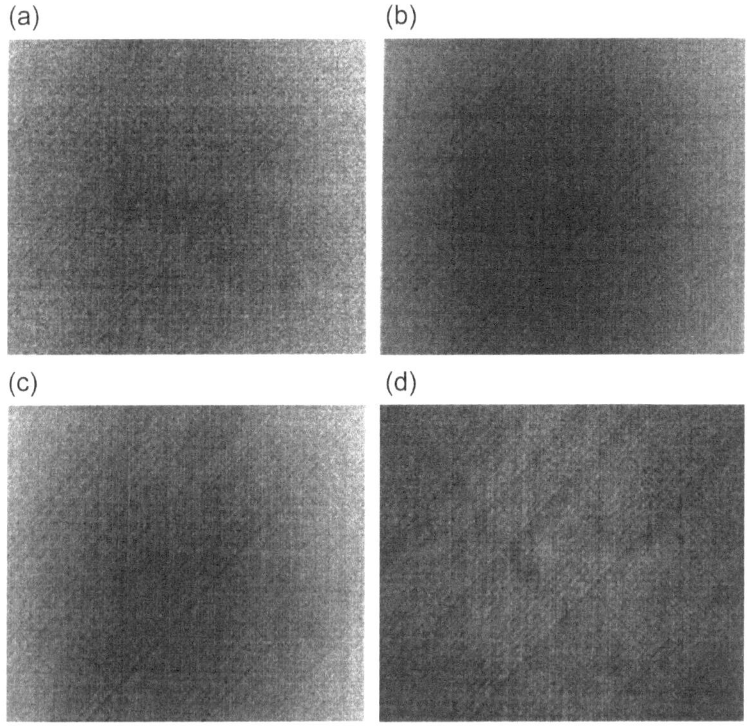

Figure 4.22 Phase images of specimen CFRP-D. (a) $f = 0.87$ Hz; (b) $f = 0.65$ Hz; (c) $f = 0.47$ Hz; (d) $f = 0.30$ Hz.

occurs when a crack, initiated at a specific point, propagates to a critical size. Conversely, composites, especially carbon-reinforced ones, have better mechanical performance; significant cracks do not develop below 60% of static failure stress [47]. This has led in the past not to consider fatigue of concern in composite design [47,48]. However, designers of composites have to face other problems. For example, composites are prone to formation of manufacturing defects that may affect their optimum performance.

Besides, the main weakness of composites is their vulnerability to impact damage. It is well known that thermoset-matrix composites like carbon/epoxy or glass/epoxy often react to the impact without visible signs while the material undergoes important internal damage that, if it remains undetected, may lead to catastrophic consequences. Lock-in thermography allows for ascertaining whether a structure has undergone any damage and, in the affirmative case, to follow the damage through the thickness. As an example,

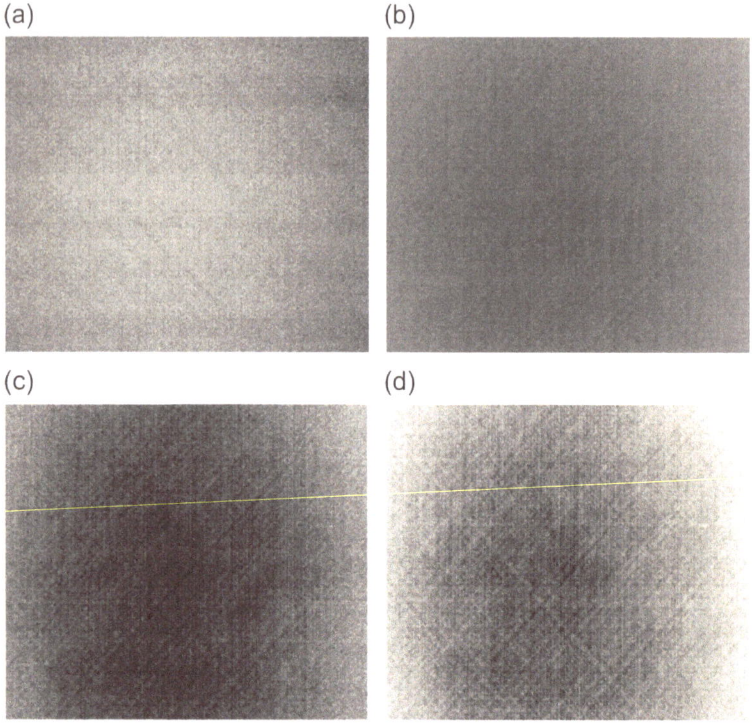

Figure 4.23 Phase images of specimen CFRP-E. (a) $f = 0.87$ Hz; (b) $f = 0.65$ Hz; (c) $f = 0.47$ Hz; (d) $f = 0.30$ Hz.

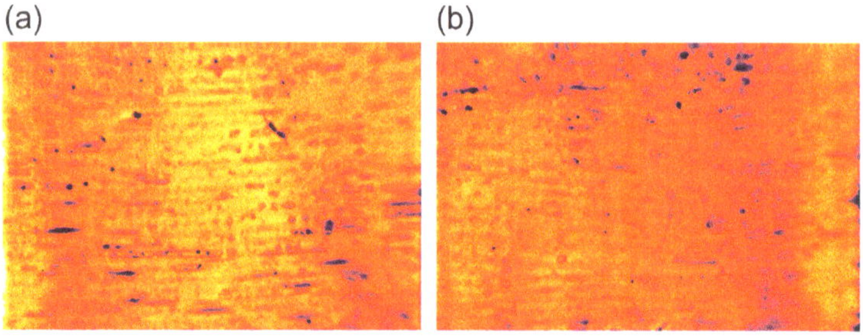

Figure 4.24 Phase images of specimens GFRP taken at $f = 0.14$ Hz (a) GFRP-A; (b) GFRP-B.

Fig. 4.26 shows three phase images of an impacted glass/epoxy specimen, taken by varying the heating frequency. More specifically, the infrared camera (SC6000) views the impacted side. The first image (Fig. 4.26(a)) at the highest frequency displays the superficial indentation (light central zone)

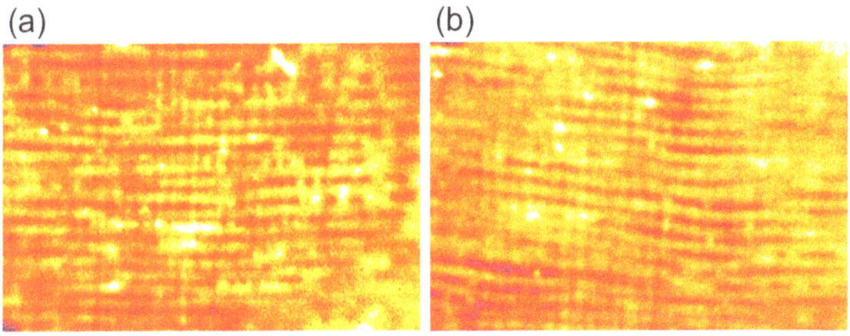

Figure 4.25 Phase images of specimens GFRP taken in transmission mode. (a) GFRP-A; (b) GFRP-B.

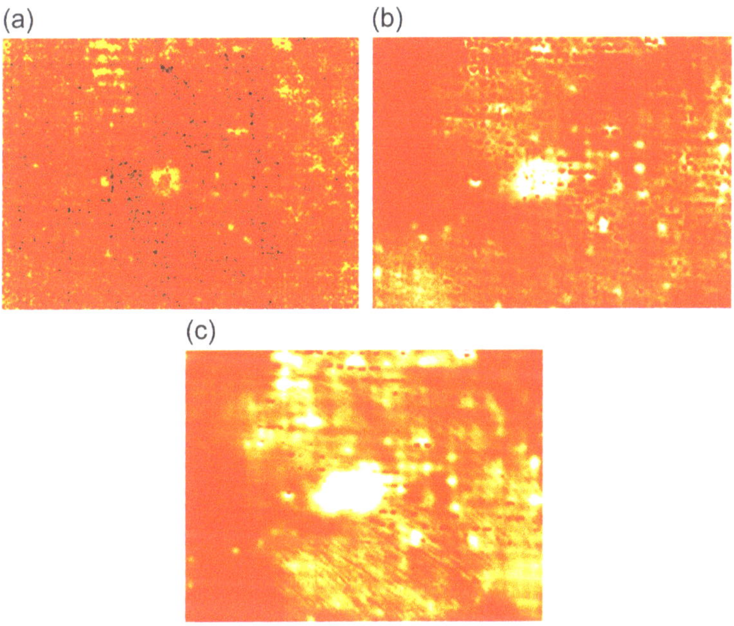

Figure 4.26 Phase images of impacted GFRP specimen. (a) $f = 0.67$ Hz; (b) $f = 0.14$ Hz; (c) $f = 0.045$ Hz.

and some diffuse porosity (light scattered spots); as f is reduced and the thermal wave reaches a deeper layer, certain of the scattered spots move elsewhere and the lighter central area enlarges (Fig. 4.26(b)) accounting for some important damage there. Such a central light zone widens up to a further reduction of f (Fig. 4.26(c)). Then, by knowing the spatial resolution of the used instrument (infrared camera and lens), it is possible to measure, from the phase images, the extension of the damaged area so as to reconstruct the damage progress through the laminate thickness [46]. Of course, the progress of the porosity distribution through the thickness is automatically accounted for.

In Fig. 4.27 for a carbon/epoxy specimen, an example of through-the-thickness reconstruction of the impact damage progress is shown. Starting from the top, it is possible to see as the damaged area A_D from an almost circular shape, which relates to the indentation, widens and develops as branched structure via the heating frequency is reduced and the depth p increases. In particular, the damaged area for the CFRP specimen (Fig. 4.27) assumes a more complex shape in comparison with that of the GFRP specimen of Fig. 4.26. Such difference is justified by considering that the GFRP specimen involves unidirectional glass fibres oriented at 0 and 90 degrees, instead the CFRP specimen entails a more complicated architecture with fibres at 0, 90, +45 and −45 degrees as well as non crimp fabrics (NCF) layers. The damaging of composites under impact occurs through complex mechanisms involving multiple matrix cracks, fibre pull-out, fibre breakage and delaminations, all mostly driven by the stacking sequence.

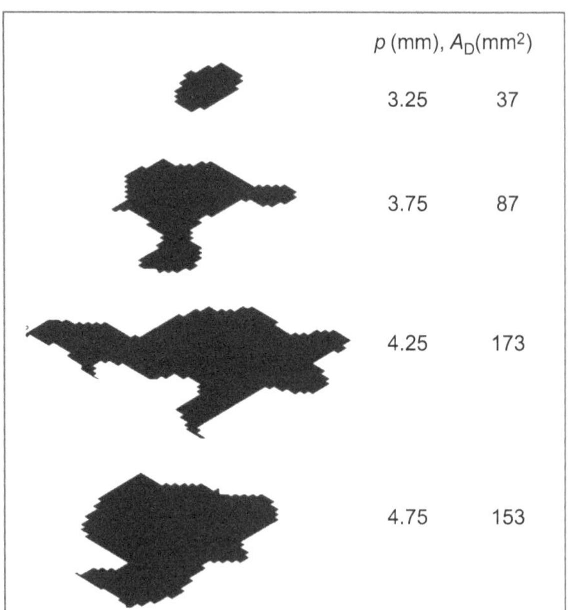

Figure 4.27 Reconstruction of the damaged area from phase images of impacted CFRP specimen.

4.5 OTHER THERMOGRAPHIC METHODS

One main requirement in PT (LPT) is the uniform distribution of the heat flux absorbed by the surface of the object under examination. This can be quite easily attained in the case of a relatively small surface, while it becomes more difficult for a large surface, which requires a certain number of evenly distributed lamps.

A technique that has some advantages over pulse thermography is the TRIR [49]; within this technique, a step heating (long pulse) is applied to the specimen while its surface temperature is monitored as a function of time. The heating pulse is long compared to the observation time and acts as a temporal step function; this technique is particularly indicated for the analysis of multilayered coatings.

In specific cases such as in the inspection of cavities and channels, when a strong heating pulse is required, the pulse thermography with *injecting water vapour* (PTJV) may be convenient. PTJV was proposed by Meola et al. [50] to detect ceramic residuals in turbine blades. The advantage of using steam is that the interface (vapour/cold wall) has a very high convective heat transfer coefficient so that the wall surface in contact with the vapour gets suddenly to a temperature very close to the vapour condensing one, thus determining a strong heat pulse.

OLT is sensitive to thermal boundaries, which do not allow for a clear and reliable discrimination between sound and defective areas. For gas-filled defects, such as cracks, or voids, this problem may be solved by selectively stimulating defective areas with acoustic waves. This is an alternative deployment of lock-in thermography, which is called *ultrasound lock-in thermography* [51]; such technique is particularly indicated for detection of corrosion and vertical cracks (perpendicular to the viewed surface). More generally, IRT associated with the detection of thermal energy produced by mechanical vibration is called *vibrothermography*; under mechanical vibration, heat is released by friction where discontinuities, such as cracks and delaminations, are located [52]. Vibrothermography [53] is also known as *stress pattern analysis by the measurement of thermal emission*, which is a tool to obtain full-field noncontact stress data and is particularly useful for metallic structures.

In addition, eddy current is used for heating coupled with lock-in thermography [54] as induced lock-in thermography and with pulse thermography [55] as pulse eddy current. Lock-in thermography is also used in conjunction with the electronic speckle-pattern interferometry and is called optical lock-in interferometry [56].

In the presence of defects buried inside a body of complex geometry and of thermal characteristics too close to the sound material, a viable alternative to the basic OLT approach may be the *rear heating phase thermography* (RHPT) where the transmitted part of the thermal wave is exploited for defects detection [57]. This approach allows for close-up view and, so, also for inspection of small parts.

A technique that combines the advantages of both pulse and lock-in thermography without sharing their drawbacks is the *pulse phase thermography* (PPT) [58]. Within PPT, the component is pulse heated as in PT and the mix of frequencies of the thermal waves launched into the specimen is unscrambled by performing the Fourier transform of the temperature evolution over the field of view; results can be presented as in LT in terms of phase or magnitude images. Deeper layers can be inspected with less influence of surface infrared and optical features. It was introduced by Maldague and Marinetti in 1996 [58] and later analysed by Vavilov and Marinetti [59]. Within PPT, the specimen is pulse heated as in PT and the mix of frequencies of the thermal waves launched into the specimen is unscrambled by performing the Fourier transform of the temperature evolution over the field of view; results are presented in terms of phase or magnitude images as in LT. Deeper thicknesses can be inspected with less influence of surface infrared and optical features. A comparison between PT, LT and PPT applied to masonry structures was performed by Carlomagno and Meola [60] with the outcome that the PPT technique may be successful in the detection of buried anomalies but may fail to get the correct depth at which any anomaly is located.

4.6 SOME APPROACHES TO APPLICATION IN FIELD

Nondestructive methods may find two types of application:

- assure a part is free of defects before final assembly
- check the conditions of a part in service.

The first application is the simplest one and resembles a laboratory test with the difference that the specimen is replaced by a real part. The application in field poses some problems involving amongst others the portability of the equipment and the interference of the surrounding environment. Indeed, the availability of systems for in-field and for in-process inspection was always a primary concern to the industrial enterprise. In this context, a noncontact and remote operation infrared imaging device offers a great advantage. For this reason, IRT became attractive since the creation of the first prototype in the 1960s; however, at that time, the system configuration was composed of different items and was cumbersome to carry and manage [61]. Then, the first step toward application in field was a compact system, easy to carry [62].

Now, due to great technological progress, infrared devices are quite light and easy to carry and the attention is mainly turned toward the control in process, such as for the case of an automated composite material layup [63] or for the inspection outside in a field environment where electrical power is unavailable. For these cases a thermal heat source was configured to be removable coupled to the material to generate a localized thermal field [64]; such a heat source consists of an exothermic chemical reaction to be activated under request. Another way to obtain thermal signatures from anomalies, over large parts, is by passing electric current through the part [65], which causes ohmic heating in sites containing dislocations. A viable alternative is the exploitation of an electromagnetic field, which induces eddy current [66] and which can be coupled with a handheld infrared system [67]. Attention was also turned toward portable imaging apparatus coupled with a sound source for field inspection on in-service aircraft fuselage and structural components [68,69]. An inspection system that combines IR imaging and laser-ultrasound results in a 3D projection of complex shape composites and is described in patent US7605924 [70]. The theoretical approach is based on the production of sound by thermal expansion on the surface struck by a laser beam; the use of a portable laser ultrasound system makes possible in-field inspection as disclosed by the patent US7612894 [71].

CONCLUSIONS TO CHAPTER 4 AND INTRODUCTION TO CHAPTER 5

In Chapter 4, the approaches for NDT and evaluation with IRT were illustrated with some key examples. In particular, the capability of IRT to detect many types of defects in composites, as well delamination and impact damage, has been shown. However, within the materials science context, the value of IRT is not limited to nondestructive evaluation, as for many NDT techniques, but it crosses into other areas such as monitoring of fatigue tests and loading events. Chapter 5 will be devoted to the use of IRT to monitor loading events such as bending and impact tests.

REFERENCES

[1] V. Vavilov, Thermal nondestructive testing: short history and state-of-art, in: D. Balageas, G. Busse, G.M. Carlomagno (Eds.), Proceedings of the QIRT92, Eurotherm Series 27. EETI ed., 1992, pp. 179–193.
[2] R.D. Parker, Thermic Balance of Radiometer, US1099199, 1914.
[3] G.A. Barker, Apparatus for Detecting Forest Fires, US1959702, 1934.
[4] G.T. Nichols, Temperature Measuring, US2008793, 1935.
[5] W.S. Beller, Navy sees promise in infrared thermography for solid case checking, Missiles Rochets 16 (1965) 22.
[6] D.R. Green, An instrument for non-destructively testing fuel core to cladding heat transfer, Nucl. Sci. Eng. 12 (1962) 271–277.
[7] D.R. Green, Principles and applications of emittance-independent infrared non-destructive testing, Appl. Opt. 7 (1968) 1779–1786.
[8] ASTM E2582-07, Standard Practice for Infrared Flash Thermography of Composite Panels and Repair Patches Used in Aerospace Applications, 2007.
[9] H.S. Carslaw, J.C. Jaeger, Conduction of Heat in Solids, Clarendon Press, Oxford, UK, 1959.
[10] C. Cattaneo, A form of heat conduction equation which eliminates the paradox of instantaneous propagation, Comptes Rendues 247 (1958) 431–433.
[11] C. Meola, G.M. Carlomagno, Application of infrared thermography to adhesion science, J. Adhes. Sci. Technol. 20 (2006) 589–632.
[12] C. Meola, G.M. Carlomagno, Recent advances in the use of infrared thermography, Meas. Sci. Technol. 15 (2004) R27–R58.
[13] X. Maldague, Introduction to NDT by active infrared thermography, Mater. Eval. 6 (2002) 1060–1073.
[14] C. Meola, G.M. Carlomagno, L. Giorleo, Geometrical limitations to detection of defects in composites by means of infrared thermography, J. Nondestruct. Eval. 23 (4) (2004) 125–132.
[15] V.P. Vavilov, D.P. Almond, G. Busse, E. Grinzato, J.-C. Krapez, X. Maldaque, S. Marinetti, W. Peng, V. Shirayev, D. Wu, Infrared thermographic detection and characterization of impact damage in carbon fibre composites: results of the round robin test, in: Proc. QIRT 98, Akademickie Centrum Graficzno-Marketingowe, Łódź, Poland, 1998, pp. 43–52.
[16] S.M. Shepard, Flash thermography of aerospace composites, in: Proceedings of the 4th Pan American Conference for NDT, Buenos Aires, Argentina, October 2007 papers 132.
[17] A. Boritu, V. Anghel, N. Constantin, M. Găvan, A. Pascu, Non-destructive inspection of composite structures using active IR-thermography methods, UPB Sci. Bull. Series D 73 (1) (2011) 71–84.
[18] H.I. Ringermacher, R.J. Archacki, W.A. Veronesi, Non-destructive Testing: Transient Depth Thermography, US5711603, 1998.
[19] J.G. Sun, Method for Determining Defect Depth Using Thermal Imaging, US6542849, 2003.
[20] J.G. Sun, Analysis of pulsed thermography for defect depth prediction, J. Heat Transfer 128 (2006) 329–338.
[21] W.J. Parker, R.J. Jenkins, C.P. Butler, G.L. Abbott, Flash method of determining thermal diffusivity, heat capacity, and thermal conductivity, J. Appl. Phys. 32 (1961) 1679–1684.
[22] ASTM E1461-01, Standard Test Method for Thermal Diffusivity of Solids by the Flash Method ASTM International, West Conshohocken, PA, 2001.
[23] H.I. Ringermacher, D.R. Howard, R.K. Pandey, Transient Thermography Measurement of a Metal Layer Thickness, EP1173724, 2002.

[24] C. Meola, C. Toscano, Flash thermography to evaluate porosity in carbon fiber reinforced polymer (CFRPs), Materials 7 (3) (2014) 1483–1501.
[25] G. Hendorfer, G. Mayr, G. Zauner, M. Haslhofer, R. Pree, Quantitative determination of porosity by active thermography, in: D.O. Thompson, D.E. Chimenti (Eds.), Review of Quantitative Non-destructive Evaluation, vol. 26, 2007, pp. 702–708.
[26] G. Mayr, G. Handorfer. Porosity determination by pulsed thermography in reflection mode, in: Proceedings of the 10th International Conference on Quantitative InfraRed Thermography, paper QIRT2010-933 Quebec Canada, 2010.
[27] G. Mayr, B. Plank, J. Sekelja, G. Hendorfer, Active thermography as a quantitative method for non-destructive evaluation of porous carbon fiber reinforced polymers, NDT&E Int. 44 (2011) 537–543.
[28] G. Wrobel, Z. Rdzawski, G. Muzia, S. Pawlak, Determination of thermal diffusivity of carbon/epoxy composites with different fiber content using transient thermography, J. Achiev. Mater. Manuf. Eng. 37 (2009) 518–525.
[29] H. Bhatt, K.Y. Donaldson, D.P.H. Hasselman, K. Chyung, M.P. Taylor, Effect of fibre orientation on the thermal conductivity of a uniaxial carbon fibre-reinforced aluminoborosilicate glass-matrix composite for various specimen geometries, J. Mater. Sci. Lett. 10 (1991) 1267–1270.
[30] C. Toscano, C. Meola, M.C. Iorio, G.M. Carlomagno, Porosity and inclusion detection in CFRP by infrared thermography, Adv. Opt. Technol. (2012) 6.
[31] C. Toscano, C. Meola, G.M. Carlomagno, Porosity distribution in composite structures with infrared thermography, J. Compos. 1–8 (2013) 2013.
[32] G.M. Carlomagno, P.G. Berardi, Unsteady thermotopography in non-destructive testing, in: C. Warren (Ed.), Proc. III Infrared Inform. Exch, AGA, St. Louis, MO, 1976, pp. 33–40.
[33] G. Busse, Optoacoustic phase angle measurement for probing a metal, Appl. Phys. Lett. 35 (1979) 759–760.
[34] A. Letho, J. Jaarinen, T. Tiusanen, M. Jokinen, M. Luukkala, Magnitude and phase in thermal wave imaging, Electron. Lett. 17 (1981) 364–365.
[35] J.-L. Beaudoin, E. Merienne, R. Danjoux, M. Egee, Numerical system for infrared scanners and application to the subsurface control of materials by photothermal radiometry, Proc. SPIE 590 (1985) 287–292.
[36] G. Busse, D. Wu, W. Karpen, Thermal wave imaging with phase sensitive modulated thermography, J. Appl. Phys. 71 (1992) 3962–3965.
[37] R.L. Thomas, P.K. Kuo, L.D. Favro, Vector Lock-in Imaging System, US4878116, 1989.
[38] G. Busse, Patent, DE4203272, 1992.
[39] C. Meola, G.M. Carlomagno, A. Squillace, G. Giorleo, Non-destructive control of industrial materials by means of lock-in thermography, Meas. Sci. Technol. 13 (2002) 1583–1590.
[40] S. Huth, O. Breitenstein, A. Huber, D. Dantz, U. Lambert, F. Altmann, Lock-in IR-thermography – a novel tool for material and device characterization, Solid State Phenom. 82–84 (2001) 741–746.
[41] J.C. Krapez, Compared performances of four algorithms used for modulation thermography, in: D. Balageas, G. Busse, C. Carlomagno (Eds.), Proc. 4th Conference on Quantitative InfraRed Thermography – QIRT, Eurotherm Seminar 60, Lodz, Pologne, September 7–10, 1998, pp. 148–153.
[42] C. Ibarra-Castanedo, M. Genest, J.M. Piau, S. Guibert, A. Bendada, X.P.V. Maldague, Active infrared thermography techniques for the non-destructive testing of materials, in: C.H. Chen (Ed.), Chapter XIV of the Book: "Ultrasonic and Advanced Methods for Non-destructive Testing and Material Characterization", 2007, pp. 325–348.

[43] O. Breitenstein, W. Warta, M. Langenkamp, Lock-in Thermography: Basics and Use for Evaluating Electronic Devices and Materials, Springer, 2nd. s.l., 2010.
[44] A.C. Loos, G.S. Springer, in: C.E. Browning (Ed.), Composite Materials, Quality Assurance and Processing, ASTM STP 797, 1983, pp. 110−118.
[45] L. Liu, B.M. Zhang, D.F. Wang, Z.J. Wu, Effects of cure cycles on void content and mechanical properties of composite laminates, Compos. Struct. 73 (2006) 303−309.
[46] C. Meola, G.M. Carlomagno, Infrared thermography to evaluate impact damage in glass/epoxy with manufacturing defects, Int. J. Impact Eng. 67 (2014) 1−11.
[47] M. Hiley, Fractographic Aspects of Fatigue Failure in Composite Materials, GARTEUR AG20 DERA/MSS/MSMA2/TR00168/GRATEUR Final Report TP112, 2001.
[48] B. Harris, Fatigue in Composites, Science and Technology of the Fatigue Response of Fibre Reinforced Plastics, Woodhead Publishing, 2003.
[49] L.C. Aamodt, J.W. Maclachlan Spicer, J.C. Murphy, Analysis of characteristic thermal transit times for time-resolved infrared radiometry studies of multilayered coatings, J. Appl. Phys. 68 (1990) 6087−6098.
[50] C. Meola, G.M. Carlomagno, M. Di Foggia, O. Natale, Infrared thermography to detect residual ceramic in gas turbine blades, Appl. Phys. A 91 (2008) 685−691.
[51] A. Dillenz, T. Zweschper, G. Riegert, G. Busse, Progress in phase angle thermography, Rev. Sci. Instrum. 74 (2003) 417−419.
[52] L.H. Tenek, E.G. Henneke, Flaw dynamics and vibrothermographic thermoelastic NDE of advanced composite materials, in: Proc. Thermosense XIII, SPIE, 1467, 1991, pp. 252−263.
[53] J. Rantala, D. Wu, G. Busse, Amplitude modulated lock-in vibrothermography for NDE of polymers and composites, Res. Nondestruct. Eval. 7 (1996) 215−228.
[54] G. Riegert, T. Zweschper, G. Busse, Lock-in thermography with eddy-current excitation, Quant. Infrared Thermogr. J. 1 (2004) 21−32.
[55] I.Z. Abidin, G.Y. Tian, J. Wilson, S. Yang, D. Almond, Quantitative evaluation of angular defects by pulsed eddy current thermography, NDT&E Int. 43 (2010) 537−546.
[56] G. Busse, Lockin-ESPI interferometric imaging for remote non-destructive testing, in: Proc. Vth International Workshop, Advances in Signal Processing for Non-destructive Evaluation of Materials Québec City (Canada),2−4 Aug. 2005. © X. Maldague ed., É. du CAO, 2006, ISBN 2-9809199-0-X, pp. 57−67.
[57] C. Meola, Non-destructive evaluation of materials with rear heating lock-in thermography, IEEE Sens. J. 7 (2007) 1388−1389.
[58] X. Maldague, S. Marinetti, Pulse phase infrared thermography, J. Appl. Phys. 79 (1996) 2694−2698.
[59] V.P. Vavilov, S. Marinetti, Pulsed phase thermography and fourier analysis thermal tomography, Russ. J. Nondestruct. Test. 35 (1999) 134−145.
[60] G.M. Carlomagno, C. Meola, Comparison between thermographic techniques for frescoes NDT, NDT&E Int. 35 (8) (2002) 559−565.
[61] R. Rinaldi, Chapter 2, in: C. Meola (Ed.), Infrared Thermography: Recent Advances and Future Trends, Bentham Science Publishers Ltd., 2012, pp. 29−59.
[62] K.R. Piety, B.D. Heise, R.A. Battenberg, W.T. King, Infrared Thermography System Including Mobile Unit, US5386117, 1995.
[63] J.F. Sjogren, J.A. Ritter, Real-Time Infrared Thermography Inspection and Control for Automated Composite Material Layup, WO2006/093703, 2006.
[64] M. Safai, G.E. Georgeson, J.F. Spalding, J.G. Thompson, Systems and Methods for Thermographic Inspection of Composite Structures, US7287902, 2007.
[65] J.C. Kennedy, C.T. Uyehara, J.G. Thompson, Non-destructive Testing System and Method Using Current Flow Thermography, US2004/0120383, 2004.

[66] M. Safai, G.E. Georgeson, K.D. Meredith, G.J. Clark, J.L. Duce, Electromagnetically Heating a Conductive Medium in a Composite Aircraft Component, US2008/0304539, 2008.
[67] P.J. Zombo, R.E. Shannon, M. Rothenfusser, M. Goldammer, C. Homma, J. Baumann, Hand Held Magnetic Induction Thermography System, WO2007/145677 (also US2008/0067455), 2007.
[68] R.L. Thomas, L.D. Favro, X. Han, Hand Held Sound Source Gun for Infrared Imaging of Subsurface Defects in Materials, US6593574, 2003.
[69] J.G. Thompson, C.T. Uyehara, Ultrasonic Thermography Inspection Method and Apparatus, EP1431755 (also US2004/0119019), 2004.
[70] D.R. Howard, H.I. Ringermacher, M. Dubois, T.G. Richter, T.E. Drake, Laser-ultrasound Inspection Using Infrared Thermography, US7605924, 2009.
[71] T.E. Drake, M. Dubois, Fiber Laser for Ultrasonic Testing, US7612894, 2009.

CHAPTER 5

Monitoring Load Events

5.1 INTRODUCTION

Within the materials science field, infrared technology can be used with the twofold purpose of nondestructive evaluation (NDE) and of monitoring load events. The use of infrared thermography as NDE technique has been already amply discussed in Chapter 4, therefore the attention is now turned toward monitoring load events. During dynamic load, two thermal phenomena may occur: the first one is the thermoelastic effect, which is reversible; the other one is the thermoplastic effect, which instead is irreversible. Both effects are accompanied by temperature variations whose visualization and measurement may be useful for the characterization of a material and for assessing the material performance. In particular, three types of load are analysed in this chapter:

- cyclic bending involving only thermoelastic effects
- impact involving only thermoelastic effects, or both thermoelastic and thermoplastic effects depending on the impact energy and on the material characteristics
- quasistatic bending involving formation and development of hot spots linked to creation and evolution of cracks and failure

5.2 SOME HINTS ON MATERIALS THERMAL EFFECTS

The thermoelastic effect was first conceived by Lord Kelvin (W. Thomson) in 1878 [1]. Many years later, in 1956 [2], Biot performed a thermodynamic analysis and formulated the classical thermoelastic equation. This one, under reversible and adiabatic conditions (ie, in the elastic regime and neglecting heat transfer within the body and to the environment), for isotropic materials can be written as:

$$\Delta T = -KT_{\mathrm{a}}\Delta\sigma \qquad [5.1]$$

Infrared Thermography in the Evaluation of Aerospace Composite Materials
ISBN 978-1-78242-171-9
http://dx.doi.org/10.1016/B978-1-78242-171-9.00005-X

where T_a is the absolute body temperature, $\Delta\sigma$ is the mean stress amplitude and K is the *material thermoelastic constant*:

$$K = \frac{\beta}{\rho c} \qquad [5.2]$$

with β the thermal expansion coefficient, ρ the density and c the specific heat. Eq. [5.1] relates the variation of temperature to the variation of volume. In particular, under adiabatic conditions, positive dilatation (traction) entails cooling of the material and vice versa [3]. In metals, the thermoelastic limit is generally assumed [4] as an indication for the yielding point. In orthotropic materials as fibre-reinforced polymers (FRP), Eq. [5.1] modifies as [5]:

$$\Delta T = -\frac{T_a}{\rho c_p(\beta_1 \Delta\sigma_1 + \beta_2 \Delta\sigma_2)} \qquad [5.3]$$

with β_1 and β_2 the thermal expansion coefficients along the principal material directions. Under certain conditions when the composite is made of plain weave fabric layers laid up to produce a symmetric laminate, the thermoelastic response can be considered as originating from the isotropic surface layer [6], making possible the use of the simpler Eq. [5.1].

In general, thermoelastic effects are associated with the thermoelastic stress analysis technique [3–6] whose principal purpose is to monitor the progression of damage in specific conditions, such as in the presence of notched specimens under cyclic tension-compression tests. The damage progression is associated with variations of temperature, which are quite significant and easy to measure. A crucial point is instead to deal with temperature variations that are experienced by the material during the purely elastic phase; such temperature variations are very small and difficult to measure. This may happen during cyclic bending tests or during low-energy impact tests.

The research group at the University of Naples Federico II succeeded in visualizing and measuring the thermoelastic effects that originate from either cyclic bending tests [7] or low-energy impact tests [8]. In particular, during each test, being cyclic bending, or impact, they acquired thermal images in time sequence starting before loading application and ending after force removal to consider the evolution of thermal phenomena. To account for temperature variations with respect to the ambient, they performed data analysis by subtracting the first image ($t = 0$) of the sequence (ie, the ambient temperature of the specimen surface before load) to each subsequent image so as to have a map of temperature difference ΔT:

$$\Delta T(i,j,t) = T(i,j,t) - T(i,j,0) \qquad [5.4]$$

i and j representing lines and columns of the observed surface temperature map.

5.3 CYCLIC BENDING TESTS

Bending tests are performed in different ways and by considering different setups. In particular, some preliminary tests are carried out as feasibility tests to ascertain the capability of an infrared imaging system to capture the weak signal associated with cyclic bending. Then, for quick testing a simple tool is considered, which allows for two cyclic bending test configurations as sketched in Fig. 5.1.

In the first case (Fig. 5.1(a)) the specimen is fixed at one extreme and is in contact with a punch in its central zone. Bending arises under the action

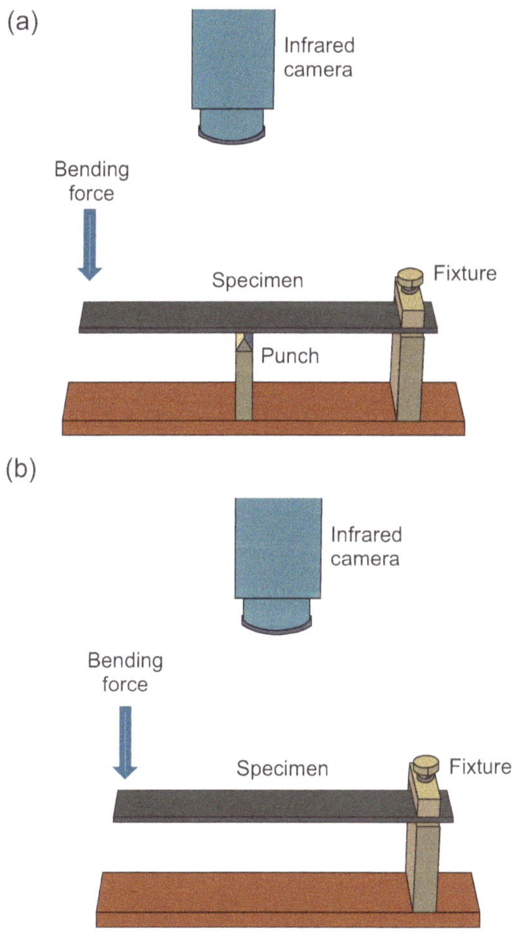

Figure 5.1 Setup for bending tests: manual force application. (a) Bending with central punch (b) cantilever beam.

of the force that is applied cyclically at the free extreme and is directed downside; the specimen displacement is applied manually. The infrared camera is positioned to see the surface on the top.

In the second case (Fig. 5.1(b)), the specimen is again fixed at one extreme but the punch is removed and then the configuration is that of a cantilever beam. Tests are performed with the force acting cyclically in the direction upside-rest, downside-rest and upside-downside (downside-upside). For each type of test the load is manually applied, while sequences of thermal images are recorded by the infrared camera. Both surfaces (top and bottom) are monitored with the infrared camera, the bottom surface being viewed with the aid of a mirror placed at 45 degrees underneath the specimen.

The main findings that have been observed are [7]:

- a sinusoidal trend that perfectly synchronizes with the cyclic displacement of the specimen free end, with decrease/increase of temperature over the surface being in tension/compression, respectively;
- a general agreement with the momentum diagram along the specimen length, while deviations of the former are observed in the presence of buried defects within the material.

The added value of Ref. [7] resides in the possibility to gain, in a fast and easy way, through the use of a simple hand-held mechanism and an infrared-imaging device, information useful for the material characterization. However, this study represents only preliminary research; in fact, the manual application of the cyclic load poses some problems such as the impracticality to change the load frequency as well as to assure overall test accuracy. This has led to the necessity of new tests with a wider variety of specimens and power-driven load application with a change of parameters such as bending amplitude and frequency.

Then, as a next step, the testing apparatus is ad hoc realized to allow for a power-driven load application with changing of the bending frequency and amplitude.

In addition, also the thermal image acquisition rate is varied in view of finding the optimal values to perform tests in a quick and reliable way [9]. It has been found that by introducing the parameter R_F as ratio between frame rate F_{Fr} and bending frequency F_B:

$$R_F = \frac{F_{Fr}}{F_b} \qquad [5.5]$$

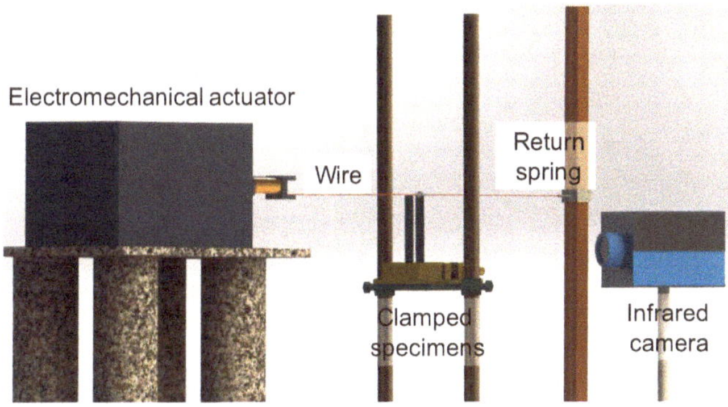

Figure 5.2 Setup for bending tests with a bending machine.

the condition:

$$R_F > 6 \quad [5.6]$$

must be satisfied to assure chasing of thermoelastic effects developing under bending tests. This condition corresponds to detecting not less than about seven frames within a single loading cycle, so as to capture average ΔT values with a certain confidence. However, the real novelty of Ref. [9] resides in the noise correction method through the reference area, which is introduced there for the first time.

The final test setup, sketched in Fig. 5.2, includes the cantilever beam specimen clamped on the bottom side (fixture) and free to bend under the cyclic harmonic force applied at the opposite end. The bending is operated with a machine that allows changing of both bending frequency and specimen deflection. More specifically, as depicted in Fig. 5.2, bending is operated with an electromechanical actuator through a wire and a return spring. The left specimen upper end is inserted into a clip attached to the wire, which forces the specimen to bend under the wire alternate displacement. The infrared camera is positioned to see one surface of the specimen [10]. In reality, as can be seen in Fig. 5.2, two specimens are clamped together on their bottom side but only one is inserted inside the clip, the second one remaining unloaded since it is being used as reference.

5.3.1 Correction of Detector Noise

A detailed description of detector noise, noise effects on the measurements of thermoelastic phenomena as well as the reference area–based correction method is given in a recent paper [11]. Herein, some details are reported to enable understanding. As already mentioned, two specimens that are identical (same material and geometry) are clamped together as shown in Fig. 5.2; one (inserted in the clip) undergoes bending, while the other one remains unloaded and is used as a reference to perform signal noise corrections. The signal that is going to be used as a reference is averaged in an area of the unloaded specimen, which consists of 20 × 40 pixels.

To show the effectiveness of the reference area over images taken with the LWIR QWIP SC6000 camera, a raw signal (a), the pure detector noise (b) and the corrected signal (c) are compared in Fig. 5.3. These graphs refer

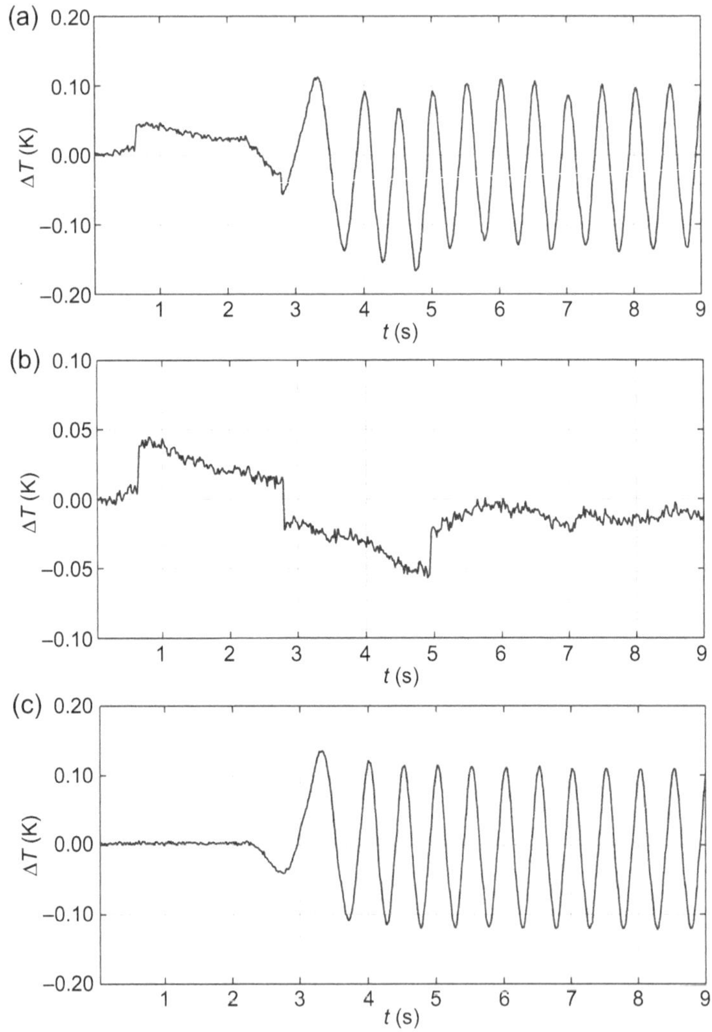

Figure 5.3 Distribution of raw ΔT_R signal (a), detector noise ΔT_N (b) and corrected ΔT_C for a Glare specimen for $F_B = 2$ Hz and $D_F = 15$ mm.

to cyclic bending tests carried out on a specimen made of Glare for $F_B = 2$ Hz and deflection $D_F = 20$ mm. Someone may advance the frequency correction; indeed, it is fully demonstrated in Ref. [11] that the frequency correction does not eliminate the alternate jump effect that, as already discussed, is due to the random detector noise as well to the camera ADC. This effect can be eliminated only through correction by the use of an unloaded reference area. In fact, the reference area, which is not interested by bending load, allows accounting for any noise-induced time variations. Then, the corrected ΔT_C is obtained as:

$$\Delta T_C = \Delta T_R - \Delta T_N \quad [5.7]$$

with ΔT_R the raw signal and ΔT_N the temporal noise measured in the unloaded reference area. To better show the effectiveness of the reference area—based correction method, it is applied to a more noisy signal coming out from bending tests on the same Glare specimen but loaded at the lower bending frequency $F_B = 0.1$ Hz. Then, the raw signal (a), the pure detector noise (b) and the corrected signal (c) are compared in Fig. 5.4. Reducing the bending frequency entails decrease of the associated ΔT; of course, the smaller the signal the greater is the influence of the noise as can be seen by comparing Fig. 5.4(a) to Fig. 5.3(a). By comparing Fig. 5.4(b) to Fig. 5.3(b), even with differences in the ΔT scale and in the recording time, it is possible to recognize differences in the jumping noise trend, which is normal because of the random nature of such a type of noise. Incredibly, the correction method is still able to recover the sinusoidal trend as can be seen by comparing Fig. 5.4(c) to Fig. 5.4(a).

5.3.2 Thermoelastic Phenomena and Bending Moment Diagram

The ΔT trend like those shown in Figs 5.3(c) and 5.4(c) can be further analysed to getting more information that may be useful for material characterization. This is done by postprocessing of the sequences of thermographic images in the Matlab environment with routines ad hoc developed. The temperature amplitude strongly depends on the position over the specimen length. Then, as sketched in Fig. 5.5 for a cantilever beam, some measurement positions (P_M) are chosen, evenly distributed along the specimen length L, starting about 10 mm far from the fixture ($x = 0$) and moving toward the forced side. In each P_M, a ΔT data plot in time, similar to those shown in Figs 5.3 and 5.4, is first extracted and corrected for noise as already described. Then, from such ΔT plots an

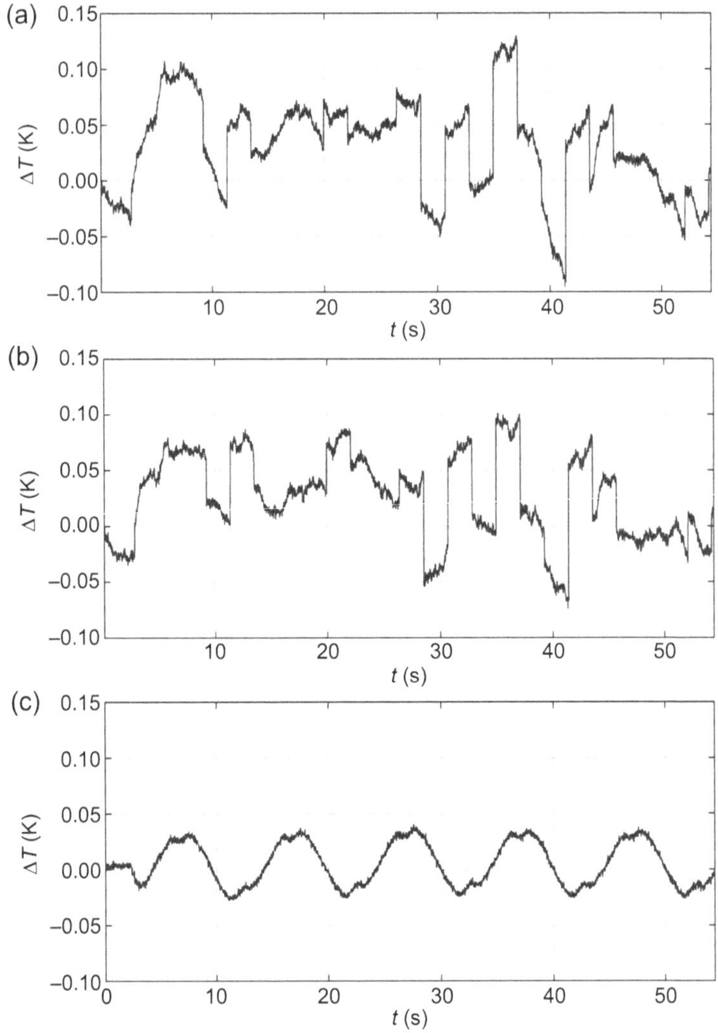

Figure 5.4 Distribution of raw ΔT_R signal (a), detector noise ΔT_N (b) and corrected ΔT_C for a Glare specimen for $F_B = 0.1$ Hz and $D_F = 15$ mm.

average signal amplitude is evaluated amongst a fixed number of cycles and named ΔT_a (the mean value in an area 20 pixels long in the x-direction and 40 pixels wide along the y-axis), by suppressing, of course, border appraisals.

The distribution of ΔT_a values over the specimen surface is compared with the bending moment diagram for the cantilever beam in Fig. 5.6 and for the configuration involving a central punch in Fig. 5.7. As can be seen

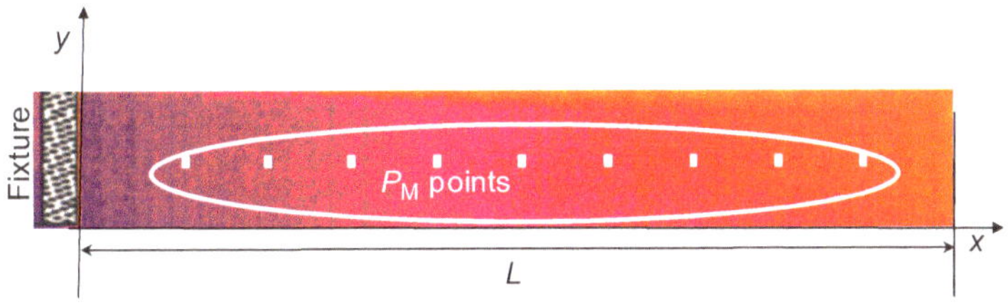

Figure 5.5 Measurement points along the specimen surface.

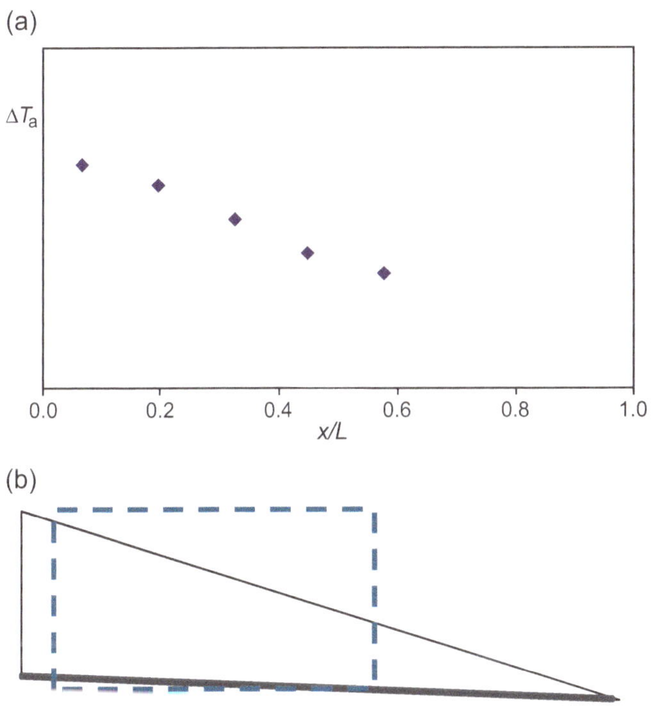

Figure 5.6 Comparison between ΔT distribution and bending moment diagram for a cantilever beam. (a) ΔT_a distribution over the specimen surface. (b) Bending moment diagram.

for both configurations a general agreement is found. It is worth noting that, while the trend remains the same, quantitative data (ΔT_a) depends on the material; then, no scale is added to Figs 5.6 and 5.7. The diagram of Fig. 5.7 has to be considered as flipped horizontally over the sketch of Fig. 5.1. Important to note is that an infrared imaging device is capable of visualizing the temperature variations induced by any change in the clamping force and loading way.

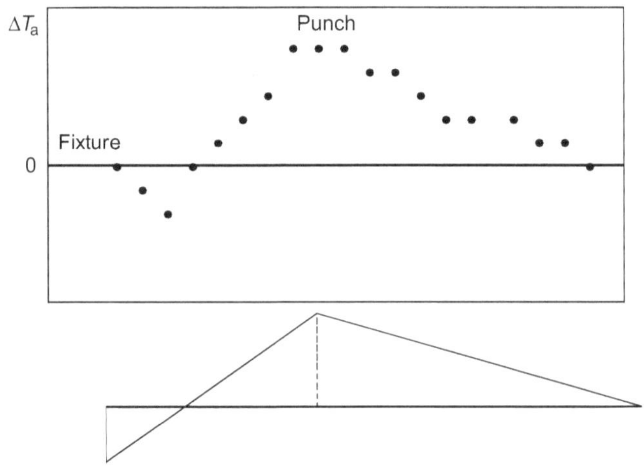

Figure 5.7 Comparison between ΔT_a distribution and bending moment diagram in presence of a central punch (Fig. 5.1).

Owing to the general agreement between ΔT_a distribution and bending moment diagram, it is possible to infer that some material characteristics can be derived by simply monitoring the thermoelastic effects. As an example, the behaviour under bending of some composites involving change of matrix and of reinforcement is herein analysed. In particular, some specimens are made of a thermoset matrix (epoxy resin) reinforced with unidirectional glass fibres. Other specimens are instead made of a thermoplastic matrix (polypropylene) reinforced with either woven glass fibres or woven jute fibres. Some jute-reinforced specimens include a modified polypropylene with addition of a certain percentage (2% or 5%) of compatibilizing agent. A type of Glare involving three aluminium layers interspersed by two layers of glass/epoxy is also used. Some details of the investigated specimens in terms of code, thickness and composition are summarized in Table 5.1.

A graph of ΔT_a values against x/L for four different composite materials is shown in Fig. 5.8.

It is possible to see that the trend of the bending moment diagram is retained for all the considered materials; in fact, for each type of material, data points appear aligned over a straight line. In particular, each line practically attains its maximum near the fixture and zero value moving far away from the fixture toward the free end ($x/L = 1$). More specifically, all lines seem to start from zero for $x/L = 1$ and to fan out with a different slope moving toward a maximum value at $x/L = 0$. ΔT_a seems to increase with increasing of the material thermal characteristics. In fact, amongst the

Table 5.1 List of specimens investigated under cyclic loading

Code	Thickness (mm)	Composition (matrix – reinforcement)
Glare	1.5	3 Al and 2 GFRP layers
GFRP	2.9	Unidirectional glass fibres impregnated with epoxy resin $[0_2,90_2]_s$
PPG	3.0	Neat PP – woven glass fibres [10]
PPJ	3.8	Neat PP – woven jute fibres [10]
PJC_2		Modified PP (2 wt% PP-g-MA) – woven jute fibres [10]
PJC_5		Modified PP (5 wt% PP-g-MA) – woven jute fibres [10]

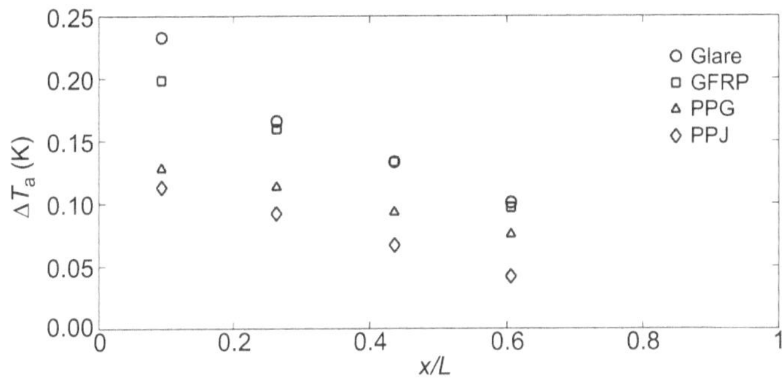

Figure 5.8 ΔT_a against *x/L* for different composite materials.

used specimens, Glare attains the highest ΔT_a and has the highest thermal diffusivity, while P_J (polypropylene composite reinforced with jute fibres) has the lowest thermal diffusivity and attains the lowest ΔT_a, notwithstanding its greater thickness.

Specimens with the presence of the compatibilizing agent in the polypropylene matrix deserve particular attention. In fact, as described in previous work [10], the presence of the compatibilizing agent induces an increase of ΔT_a; in particular, there is a systematic increase of ΔT_a with increasing of the percentage of compatibilizing agent. This effect can be seen in Fig. 5.9, which shows ΔT_a values for three types of matrix: one of pure polypropylene and two others with addition of either 2% or 5% of compatibilizing agent are compared. The variations of ΔT_a can be explained by considering that the coupling agent enhances the adhesion at the interface of composite systems. Of course, any modification at the interfacial strength would result in differences in the elastic modulus and in the thermal diffusivity, which, in turn, affect thermoelastic effects.

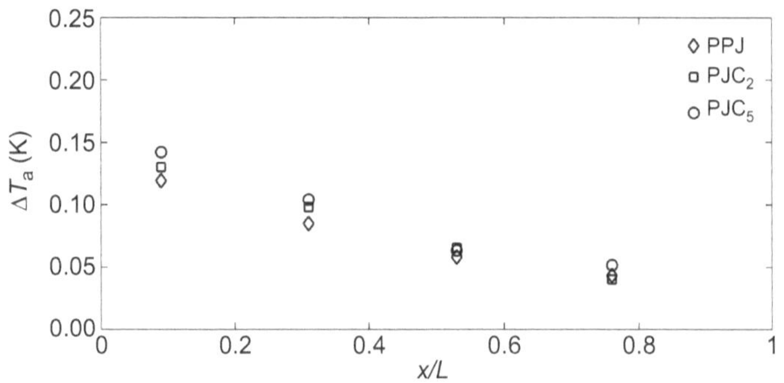

Figure 5.9 ΔT_a against *x/L* for specimens based on jute reinforcement.

5.3.3 Visualization of Secondary Effects Affecting Thermoelastic Phenomena

The temperature variations, entailed by thermoelastic phenomena linked to cyclic bending, are generally small and can be affected by secondary thermal effects. Two aspects are herein considered: one is involved with the material characteristics and the other one with the testing instrumentation and/or procedure.

With regard to the first aspect, it has been shown that a defect buried inside the material disrupts the ΔT_a trend [7]. To show such an effect, ΔT_a distributions for materials with and without defects are compared in Fig. 5.10. More specifically, the three curves refer to two carbon/epoxy specimens of which one is sound and another one presents a shallow

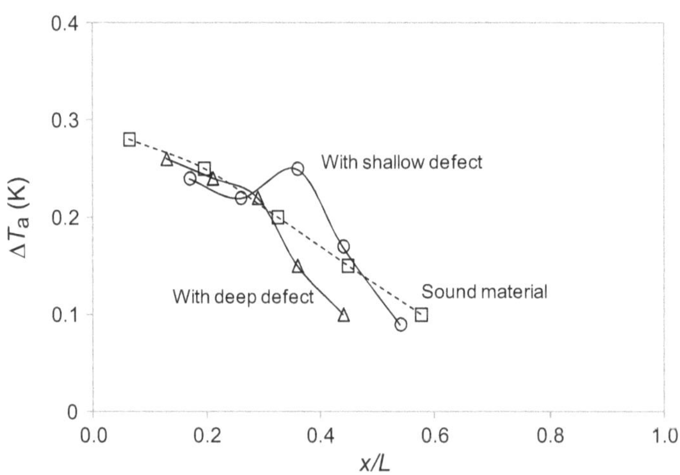

Figure 5.10 ΔT_a against *x/L* for specimens with and without defects.

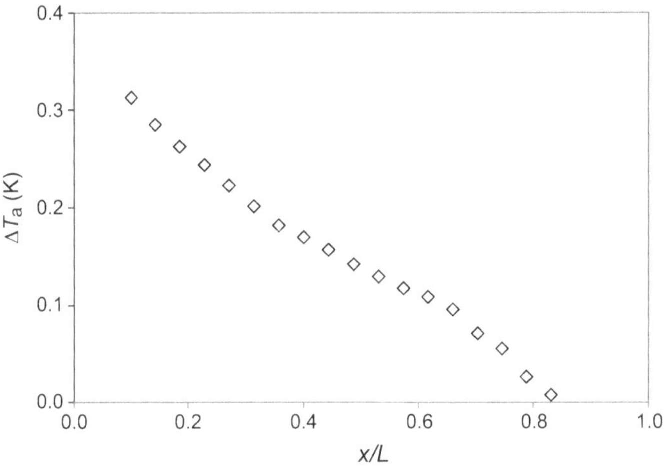

Figure 5.11 ΔT_a against *x/L* for a Glare specimen in the presence of torque.

anomaly (from one side) located at about 40% of its length. Cyclic bending tests were performed by viewing both sides with the infrared camera. As can be seen, the buried defect affects the normal ΔT_a distribution in a different way depending on its depth. At first glance it seems that the effect of the deep defect is less pronounced since no peak is present, but the change of slope is equally important. It can be inferred that the slope is driven by the flexural modulus, which is affected by the presence of a defect, independently of the depth at which such a defect is located. Instead, the ΔT_a amplitude depends on the local variation of thermal diffusivity, which depends on the local material conditions.

Owing to the second aspect, the analogy between the distribution of ΔT_a against x/L and the bending moment is valid until the applied load is pure bending. Any deviation from this can be visualized by the infrared camera. As an example, Fig. 5.11 shows how the ΔT_a distribution modifies when torque is coupled to the bending load. As can be seen, data points are no longer aligned on a straight line but on a wavy curve and, in addition, ΔT_a goes to zero at $x/L \approx 0.85$ and not at 1.0.

5.4 IMPACT TESTS

A main weakness of all simple monolithic composites is their low interlaminar strength, which makes them susceptible to impact damage. Mostly dangerous is the impact at low energy, which does not produce damage visible on the external surface but rather buried delamination between the layers. In general, composites are able to absorb the impact energy within their polymeric matrix that distributes the energy in the material; in this way, a low-energy (low-velocity) impact does not produce perforation but delamination between the layers, with no visible surface manifestation, whereas the structural integrity may be severely affected [12].

In the aircraft industry there is the habit to quantify impact threats in terms of impact energy, but, as observed by Olsson et al. [13], this is not at all correct since a small mass and a large mass impactor entails, at the same impact energy, a completely different response. A solution to this drawback was recently found by Meola and Carlomagno [14] who proposed a relationship that links the damaged area to the impact energy and to the surface area in contact with the impactor. Indeed, the impact damage of composites happens through complex mechanisms and is still not completely understood. This is mainly due to the multitude of materials that can be created by changing any of their constituents: matrix, reinforcement, stacking sequence, curing process etc. Since the introduction of composites in the construction of aircraft, a primary task was to establish the *delamination threshold load* (DTL) [15]. However, notwithstanding the huge amount of available data coming from both numerical simulation and experimental testing, a methodology to unambiguously establish the DTL still has not been completely achieved. This is because the DTL depends on many factors, first of all the material mechanical characteristics but also the geometry of the target [16,17] and of the impactor [18].

Another problem within composites is linked to defects that can be accidentally induced during their manufacturing processes. Indeed, due to the numerous parameters involved, these processes are probably primarily responsible for the occurrence of defects, particularly for porosity formation. In fact, porosity typically forms during an incorrect curing procedure due to uncontrolled or unexpected variations of the involved parameters, such as temperature, pressure, duration etc. A certain percentage of gas may remain entrapped within the material (essentially the matrix) and may give rise to formation of voids, which may modify the material in service performance. In fact, the detrimental effects of voids on composites' mechanical properties [19,20] are well recognized. Of course, the presence of porosity may also affect the behaviour of the material under impact load, with amplification of the bulk material weakness. Porosity is linked to manufacturing processes and can be reduced but not completely eliminated [21]. Perhaps this is the main reason why composite materials display a large variety of damaging behaviours under impact [22]. The negative effects of voids and other manufacturing defects on the impact damage of glass/epoxy specimens were also observed by Meola and Carlomagno [23] while using infrared thermography to investigate the response of composites to impact events. In particular, they proved that with an infrared imaging device it is possible to visualize the thermal effects that develop under low-energy impact and which may supply information about initiation and propagation of the impact damage.

Thus, once a new material is created, it is mandatory to assess its impact damage resistance. Often this feature is assessed through specific impact tests, which are intended to identify the energy that causes a delamination of a given extension. More specifically, such tests consist of impacting the laminate at a given energy, evaluating in a nondestructive way the induced damage extension and then increasing impact energies until the preset delamination has been reached. Of course, this operation is time-consuming and, sometimes, not very accurate depending on the surface under exam and on the employed instrument. In fact, despite the many nondestructive testing techniques available today, there is no certainty of having detected the actual extent of delamination. This is because the two surfaces, in a delaminated zone, tend to adhere once the impactor moved away and the resulting discontinuity is too small to affect the signal to be exploited for nondestructive testing. As demonstrated by the research group at the University of Naples, an infrared imaging device can be advantageously exploited to monitor thermal effects evolving on the surface of composite materials when subjected to impact [8,14,23]. This is because the thermal signatures can be related to what occurred to the material under load and help to understand the impact damage mechanisms as well to estimate the overall affected impact zone.

5.4.1 Test Setup and Procedure

Impact tests are performed with a modified Charpy pendulum, which allows the infrared camera to view the surface opposite to the impact as sketched in Fig. 5.12. The specimen is positioned inside a specific lodge, which is made of two large plates, each having a window to allow for contact with the hammer head from one side and optical view (by the infrared camera) from the other one. Again, as already specified for cyclic bending tests (see Section 5.3.1), a reference specimen is used for correction of the detector noise. More specifically, as shown in Fig. 5.12, a smaller

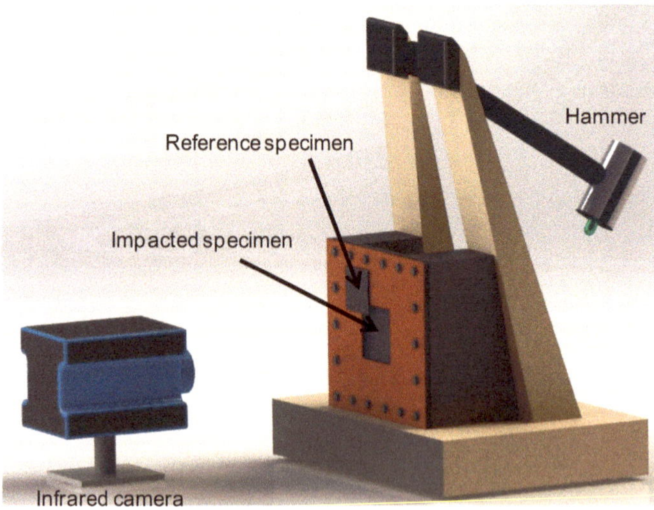

Figure 5.12 Setup for impact tests with a Charpy pendulum.

sample of the same material is attached over the specimen lodge so as to be included in the camera field of view without undergoing percussion. The impact energy is set by suitably adjusting the falling height of the hammer centre of gravity. The infrared camera acquires sequences of thermal images during the impact event.

5.4.2 Some Key Examples

Throughout the past few years, the application of infrared thermography for monitoring of impact tests was amply investigated at the University of Naples by varying the type of material, the impact energy, the hammer nose diameter, the infrared camera and the acquisition frame rate as well [8,14,23–30].

For an overview of the development of thermal phenomena under impact, two videos (GFRP and PPG) taken during impact at $E = 8.3$ J are supplied for the online version of this book. More specifically the two videos refer to ΔT images created according to Eq. [5.4].

The following are the supplementary data related to this chapter:

Supplementary data related to this article can be found online at http://dx.doi.org/10.1016/B978-1-78242-171-9.00005-X.

The video identified with the code GFRP refers to a specimen based on a thermoset matrix (epoxy resin) reinforced with unidirectional E glass fibres oriented at 0 degree and 90 degrees, or better in accordance with the stacking sequence $(0_2, 90_2)_s$, with an overall thickness of 2.9 mm. The sequence of thermal images was taken with the SC6800 camera (Flir systems) at a frame rate of 960 Hz. It is possible to see that initially, before impact, the temperature is almost uniform and equal to the ambient temperature ($\Delta T = 0$). Then, a sudden variation occurs at the impact with the appearance of some dark zones due to cooling down because of thermoelastic effects; this accounts for the surface bending under the pushing impact force. Later on, some hot spots appear over a horizontal fibres pocket; in the meantime, the dark zone first enlarges and afterwards disappears and is replaced by a warmer oblong area enclosing the hot spots, which tend to coalesce with time.

As the video-coded PPG shows, the specimen made of a thermoplastic (polypropylene) matrix reinforced with woven glass fibres behaves in a different manner under impact even if at the same energy of 8.3 J. In fact, the cooling down starts with two oblique rays that evolve into a well-shaped cross with a central zone, which is initially at ambient temperature and then tends to warm up.

For a direct comparison between the two specimens, some ΔT images taken at different time instants, before, during and after impact, are reported in Fig. 5.13 for the GFRP specimen and in Fig. 5.14 for the PPG specimen.

Starting from the GFRP specimen, the first image (Fig. 5.13(a)) appears of uniform colour, which corresponds to $\Delta T = 0$. The second image

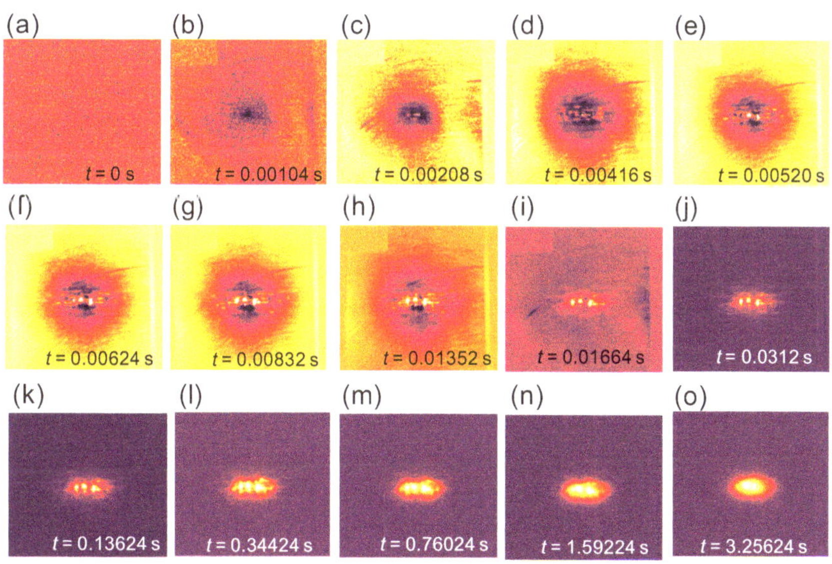

Figure 5.13 ΔT images taken at different time instants of a GFRP specimen impacted at $E = 8.3$ J.

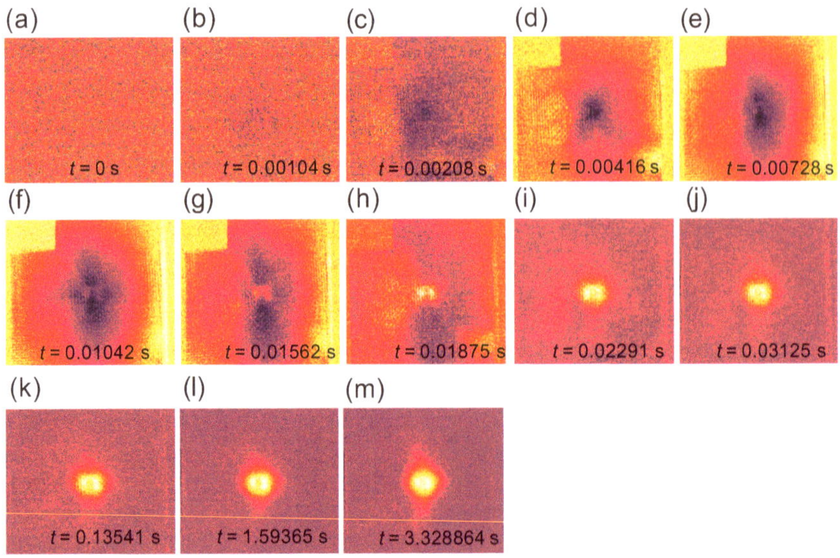

Figure 5.14 ΔT images taken at different time instants of a PPG specimen impacted at $E = 8.3$ J.

(Fig. 5.13(b) taken 1/960 s later) displays a dark zone, which, as already said, indicates the surface cooling down under the impactor pushing force. Later (Fig. 5.13(c)), some lighter (warm) tracts become visible, accounting for local heat dissipation. Indeed, as already demonstrated by Meola and Carlomagno [14], this is due to the formation of cracks and is in agreement with the general belief [31–33] that the damage process is initiated, at least in thermoset matrix-based composites, by matrix cracks that form under surface bending and in turn lead to delamination at ply interfaces. With passing time more tracts become lighter and at $t = 0.0052$ s (Fig. 5.13(e)) two hot spots appear, which are followed, as can be seen through the successive figures (Fig. 5.13(f–h)), by the appearance of other hot spots. These hot spots, which are characterized by abrupt temperature rise up to 15K (strong heat dissipation) indicate fibre breakage. Proceeding to the successive images, the negative ΔT values tend to vanish while gradually the hot spots become much milder and coalesce to form a unique warm area.

By comparing Figs 5.13 and 5.14 it is possible to better demonstrate the already cited different behaviour of the two materials of which the two specimens (GFRP and PPG) are made. In particular, as a main difference, the PPG specimen is characterized by a wider and long-lasting colder zone and a late appearance of a practically unique warm area without hot spots.

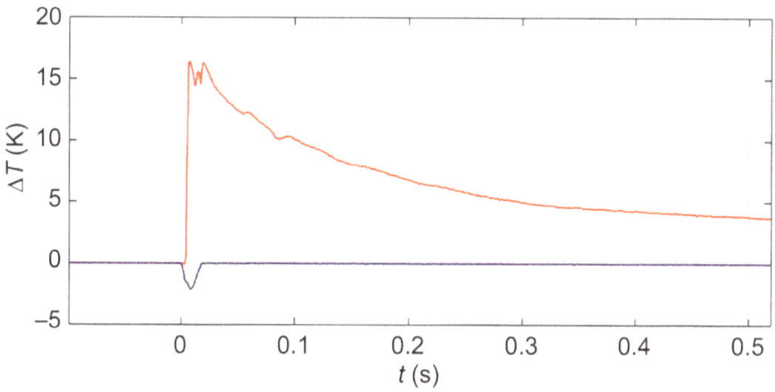

Figure 5.15 Plots of ΔT_{Max} and ΔT_{Min} for a GFRP specimen impacted at $E = 8.3$ J.

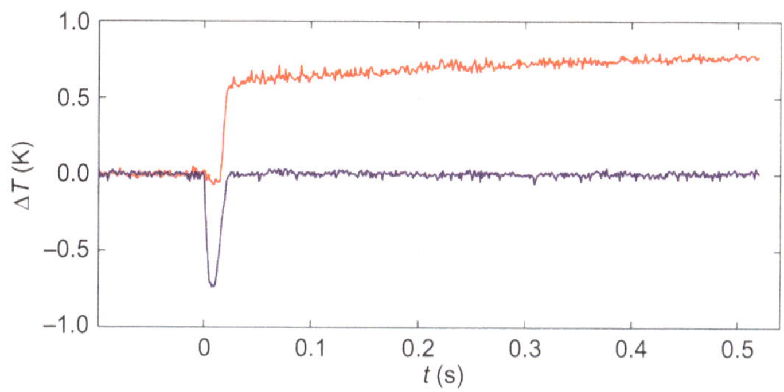

Figure 5.16 Plots of ΔT_{Max} and ΔT_{Min} for a PPG specimen impacted at 8.3 J.

For a deepening, maxima and minima ΔT values over the observed area (ΔT_{Max}, ΔT_{Min}) are extracted and plotted against time in Figs 5.15 and 5.16. As already described in Section 5.3.1, such plots are again corrected for the alternate jump noise effect by using the unloaded reference area (see Fig. 5.12).

As can be seen, the specimen GFRP displays a very high ΔT_{Max} value (over 15K) with respect to the slighter variations undergone by the PPG specimen at the same impact energy of 8.3 J. As a further observation, in the GFRP specimen, the variation of ΔT_{Max} starts almost contemporaneously with that of ΔT_{Min}; instead, in the PPG specimen, ΔT_{Max} starts with a certain delay with respect to ΔT_{Min}. This aspect is further investigated through the close view plots of Figs 5.17 and 5.18 and by considering two impact energies.

More specifically, in Fig. 5.17 plots are reported of ΔT_{Max} and ΔT_{Min} for $E = 2.8$ and 8.3 J of two GFRP specimens. Analogously, Fig. 5.18 shows

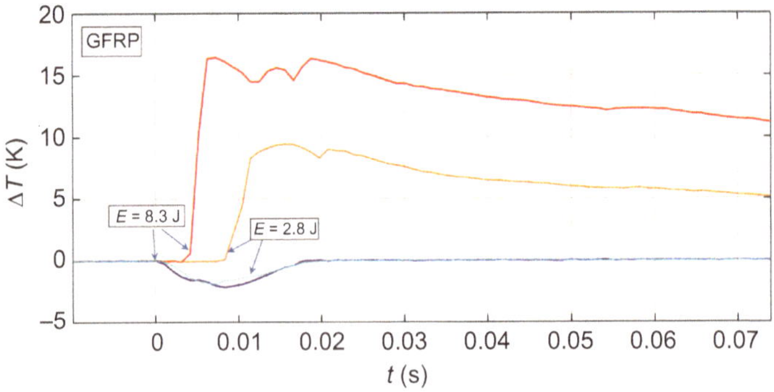

Figure 5.17 A comparison of ΔT_{Max} and ΔT_{Min} for two GFRP specimens.

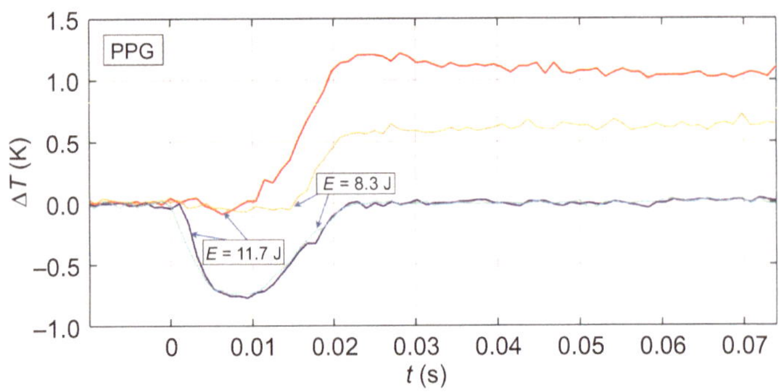

Figure 5.18 A comparison of ΔT_{Max} and ΔT_{Min} for two PPG specimens.

plots of ΔT_{Max} and ΔT_{Min} on two PPG specimens impacted at $E = 8.3$ and 11.7 J. By comparing Figs 5.17 and 5.18, several observations can be made.

First of all, for both materials, the cooling phase seems independent of the impact energy in terms of both duration and amplitude.

As already observed the cooling phase depends on the type of material; to better account for this, ΔT_{Min} values for the two specimen types are plotted together in Fig. 5.19. As can be seen, ΔT_{Min} reaches a lower value and is spiky shaped for the GFRP specimen. This is because the PPG specimen, which encloses a thermoplastic matrix, allows for larger plastic deformations with respect to the GFRP specimen enclosing a thermoset matrix. Practically, the surface of the PPG specimen deforms plastically under the impact pushing force, while the surface of the GFRP recovers its straight configuration once the contact with the impactor ceases. This different behaviour is accompanied by different heat transfer mechanisms, which affect the surface temperature distribution.

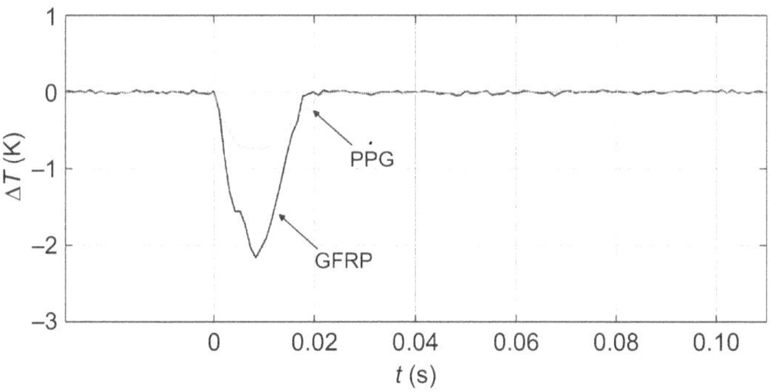

Figure 5.19 A comparison of ΔT_{Min} values for GFRP and PPG specimens impacted at $E = 8.3$ J.

For both materials, the beginning of the heating phase strongly depends on the impact energy; more specifically, there is a delay with decreasing the impact energy. In any case the ΔT amplitude is higher for the GFRP material since the thermoset matrix does not allow for large deformations, which leads to local breakage (hot spots) with, of course, dissipation of large amount of energy and in turn rise of temperature. In particular, the temperature increase is produced by the amount of energy absorbed by the material.

The mechanism of energy absorption by the composite during impact depends on many factors, such as impact velocity, geometrical parameters and inherent material characteristics (ie, brittle or ductile) [34,35]. For low-velocity impacts, the energy E_a absorbed by the specimen is generally regarded as the sum of membrane energy E_m, bending energy E_b and damage energy E_d:

$$E_a = E_m + E_b + E_d \tag{5.8}$$

The importance of each component depends on the material properties. For a brittle material, E_d includes two terms, one accounting for fibre breakage E_{db} and the other one for delamination E_{dd}. Instead, for a ductile material, the energy is predominantly spent in deformation and delamination perpendicular to the impactor axis. It has also been shown that any change in the matrix characteristics affects the material behaviour under impact [36]. In fact, it is sufficient to add 2% of a compatibilizing agent, based on maleic anhydride, to have a material that undergoes breakage at an impact of 20 J, unlike the pure polypropylene [36]. This is because the presence of the compatibilizing agent enhances the interface strength and prevents large deformations.

The importance of the impact damage can be classified owing to the ΔT_{Max} value, which is proportional to the amount of absorbed energy. Values of ΔT_{Max} for different types of composites are collected in Table 5.2. Specifically, five types of composites are considered:

- two (CFRP and GFRP) are based on a thermoset matrix with reinforcement made of either carbon (CFRP) or glass (GFRP)
- one is a hybrid composite (Glare) including three aluminium layers with two GFRP layers in between them
- one (PPG) includes a thermoplastic matrix (pure polypropylene) reinforced with glass fibres [36]
- one (PG_C) includes a thermoplastic matrix modified with the addition of 2% of a compatibilizing agent and reinforced with glass fibres [36].

From Table 5.2 the following comments can be made:

- In general, ΔT_{Max}, for any type of material, increases with increasing the impact energy.
- For the same impact energy, a thermoset matrix–based material achieves higher ΔT_{Max} values.
- It seems that, for the same thermoset matrix, higher ΔT_{Max} values are achieved in the presence of carbon fibres with respect to glass fibres.

However, some discrepancies can be observed. First of all, some GFRP specimens display higher ΔT_{Max} values at lower impact energy, or two completely different values for two impacts at the same energy (see values for $E = 9.7$ J). This happens because of the presence of manufacturing defects, like fibre misalignment and/or porosity, buried in the material [23]. In fact, the presence of defects enhances the material weakness. In particular, a local thickening in the matrix may cause premature formation of cracks and may modify the load distribution as well. In addition, a negative role is played by misalignment of fibres; it has already been observed [23] in the breakage of fibres at the tip of a bifurcation.

With regard to specimens with a thermoplastic matrix, two observations can be made. One regards the different behaviour with increasing the impact energy; in fact, for both materials PPG and PG_C ΔT_{Max} increases as E increases from 5 up to 10 J and decreases for $E = 15$ J. Such a decrease can be explained by considering that the used thermoplastic matrix composites react to the impact with visible modifications, like metals, displaying an indentation (a small concavity) on the impacted side and a protrusion on the rear one [36]. These modifications, which are very slight for very low-impact energy and become ever more pronounced by increasing the impact energy, affect the energy absorption mechanisms and, in turn, the local

Table 5.2 Maxima ΔT values for different types of composites impacted at different energies

Code	Matrix	Thickness (mm)	Fibres	E (J)	ΔT_{Max} (K)
CFRP	Thermoset	2.5	Carbon (1D 0/45/90)	0.5	1.6
CFRP	Thermoset	2.5	Carbon (1D 0/45/90)	1.2	1.9
CFRP	Thermoset	2.5	Carbon (1D 0/45/90)	2.8	7
CFRP	Thermoset	2.4	Carbon (1D 0/45/90)	8.3	25
CFRP	Thermoset	2.4	Carbon (1D 0/45/90)	9.7	19.9
CFRP	Thermoset	2.4	Carbon (1D 0/45/90)	11.7	28
GFRP	Thermoset	2.9	Glass (1D 0/90)	2.8	9.5
GFRP	Thermoset	2.9	Glass (1D 0/90)	8.3	16.4
GFRP	Thermoset	2.9	Glass (1D 0/90)	9.7	10.7
GFRP	Thermoset	2.9	Glass (1D 0/90)	9.7	24.7
GFRP	Thermoset	2.9	Glass (1D 0/90)	11.7	16.9
GFRP	Thermoset	2.9	Glass (1D 0/90)	15	10.5
GFRP	Thermoset	2.9	Glass (1D 0/90)	18.7	13.3
GFRP	Thermoset	2.9	Glass (1D 0/90)	15.9	29.5
GFRP	Thermoset	2.9	Glass (1D 0/90)	16.1	30
GFRP	Thermoset	2.9	Glass (1D 0/90)	19	26.3

Continued

Table 5.2 Maxima ΔT values for different types of composites impacted at different energies—cont'd

Code	Matrix	Thickness (mm)	Fibres	E (J)	ΔT_{Max} (K)
GFRP	Thermoset	2.9	Glass (1D 0/90)	24.9	30.7
GFRP	Thermoset	2.9	Glass (1D 0/90)	12.3	32
Glare	Thermoset	1.5	Glass (1D +45/−45)	2.7	5.6
Glare	Thermoset	1.5	Glass (1D +45/−45)	3.6	4.4
Glare	Thermoset	1.5	Glass (1D +45/−45)	4.5	5.6
Glare	Thermoset	1.5	Glass (1D +45/−45)	11.7	11.9
PPG	Thermoplastic	3.0	Glass (woven 0/90)	5	1.7
PPG	Thermoplastic	3.0	Glass (woven 0/90)	8.3	1.2
PPG	Thermoplastic	3.0	Glass (woven 0/90)	9.7	2.4
PPG	Thermoplastic	3.0	Glass (woven 0/90)	15	1.2
PPG	Thermoplastic	3.0	Glass (woven 0/90)	20	1.5
PG_C	Thermoplastic/compatibilizing agent	3.0	Glass (woven 0/90)	5	1.7
PG_C	Thermoplastic/compatibilizing agent	3.0	Glass (woven 0/90)	8.3	2.3
PG_C	Thermoplastic/compatibilizing agent	3.0	Glass (woven 0/90)	9.7	2.6
PG_C	Thermoplastic/compatibilizing agent	3.0	Glass (woven 0/90)	15	1.5
PG_C	Thermoplastic/compatibilizing agent	3.0	Glass (woven 0/90)	20	30

temperature rise. Most probably, there is a critical impact energy for which the material undergoes local thinning under deformation. Of course, this may enhance removal of the heat generated by the impact, preventing a rise in temperature. Another observation regards the behaviour for $E = 20$ J. In fact, while the PPG specimen displays ΔT_{Max} values slightly higher than those relative to $E = 15$ J, the PG_C specimen displays an abrupt temperature rise to 30 degrees accounting for material breakage. This means that the presence of the compatibilizing agent in the matrix prevents large deformations in the impact zone favouring local material breakage.

It is worth noting that Table 5.2 has been prepared with the intent to supply some general indications mainly devised to evidence the occurrence of breakage as linked to temperature rise. Of course, for a better comprehension of the material behaviour under impact, from temperature variations, it is important to consider other factors such as the distribution of the clamping force over the specimen surface during the impact, the impactor geometry and other variable testing factors.

5.4.3 Measurement of Impact-Damaged Area

Assessing the importance of the damage for a given energy represents the ultimate outcome of impact tests. However, we have to first define the meaning of the impact-damaged zone. We can distinguish almost two zones of different importance: one of severe damage including fibre breakage and the other one of slight delamination. The first one may be small, sometimes including only hot spots, or wider including breakage and important delamination. The second one is quite large, enclosing small delamination and the entire impact-affected area. The extension of overall impact-damaged area can be approximately sketched (Fig. 5.20), in analogy with the propagation of seismic waves. More specifically, the central red zone is that impacted by strong damage and is surrounded by circles of increasing diameter and decreasing damage strength.

However, this is only a general representation, the description being only qualitative and without the need of being an asymmetric one, while, for a full description, the local material characteristics must also be considered. Indeed, due to the vast variety of composites, it is quite impossible to draw a clear picture of the impact damage mechanisms. On the other hand, these mechanisms are still not completely understood. We do not claim to be exhaustive in our study but only draw attention to the possibility to obtain information useful to fill the existing gaps by exploiting infrared thermography.

Figure 5.20 Schematic representation of impact damage levels.

What we currently know is that thermoset matrix–based composites become damaged on the side opposite to that struck by the impactor, often, without any visible perception on the impacted side, or barely visible indentation [12] (Fig. 5.21). Signs of impact damage remain visible to the naked eye only in the presence of composites made of glass fibres embedded in a translucent matrix as shown in Fig. 5.22. Unfortunately, composites mostly used in aircraft construction include carbon fibres and epoxy resin with, in any case, an opaque finish that prevents seeing buried damage.

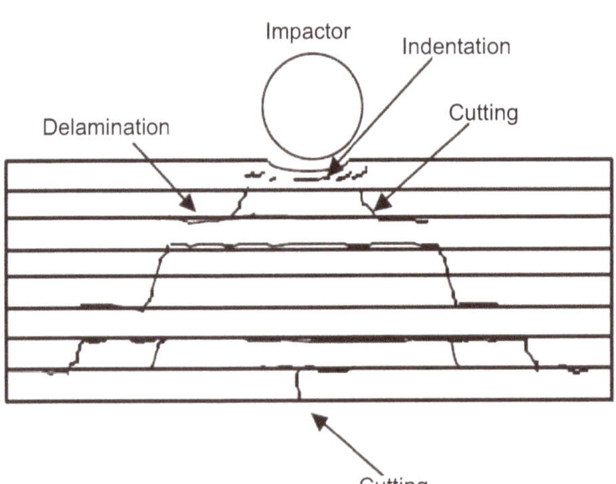

Figure 5.21 Sketch of impact-damaged thermoset matrix composites.

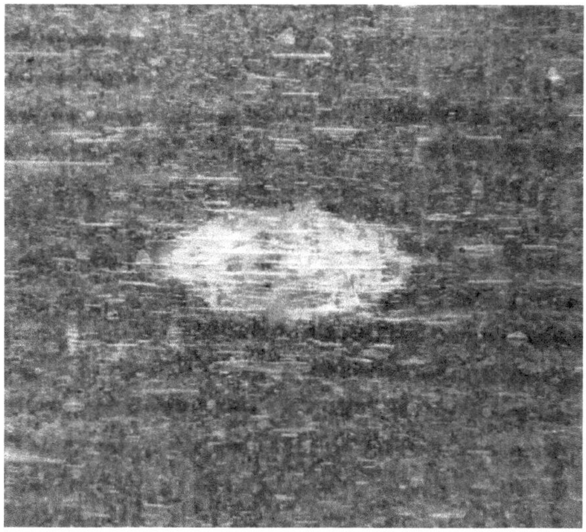

Figure 5.22 Picture of translucent glass/epoxy impacted at $E = 12$ J.

Composites including a thermoplastic matrix behave in a different manner under impact. In fact, they undergo visible deformation displaying an indentation (a small concavity) on the impacted side and a protrusion on the rear one. Of course, these modifications are very small for very low impact energy, while they become ever more evident with increasing the impact energy. However, thermoplastic composites, while resembling ductile metals for the superficial appearance, are characterized by more complex impact damage mechanisms, which strongly depend on the type of matrix and on the impact energy.

Whatever the type of composite, it is important to know:

- if a red zone (with material breakage) is present and how wide it is
- what happens to the material in the absence of breakage
- how wide the impact-affected zone is

Finding an answer to these questions may help the design of composite materials. Infrared thermography can be exploited to get such answers.

To get information about the overall extension of the impact damage, sequences of ΔT images are subjected to successive postprocessing with routines specifically developed in the Matlab environment. To locate the damage, it is important to discriminate between sound and damaged areas. Bearing in mind that we are analysing thermal images, what we can measure is the extension of the warmed-up area, which also means delineating the zone affected by temperature increase induced by the

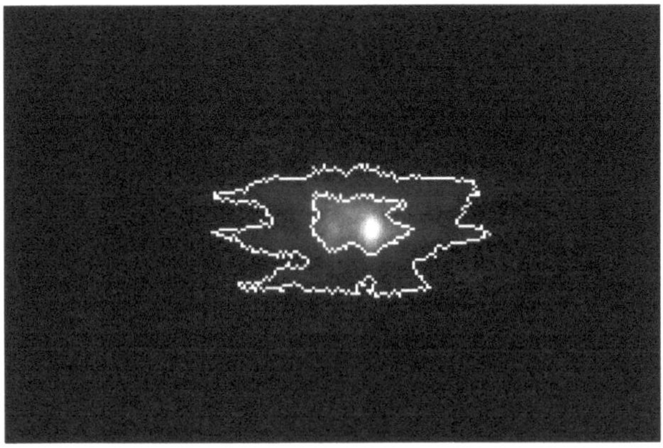

Figure 5.23 Contoured impact-damaged areas for a GFRP specimen impacted at $E = 2.8$ J.

dissipated impact energy. Then, a temperature difference threshold ΔT_T must be introduced to be considered as a limit in the ΔT map to clearly identify a zone of a certain damage level. In particular, a high ΔT_T value surrounds a zone of important damage, while a lower ΔT_T value encircles a zone of lighter damage. For ΔT_T becoming very small, the overall impact-affected zone will be contoured.

Examples of contoured impact-damaged areas are shown in Figs 5.23–5.25. In each figure, the smaller one (internal contour) is obtained with $\Delta T_T = 0.5$K and is called A_{Di}, while the larger one is obtained with $\Delta T_T = 0.1$K and is called A_{De}. In particular, Figs 5.23 and 5.24 refer to

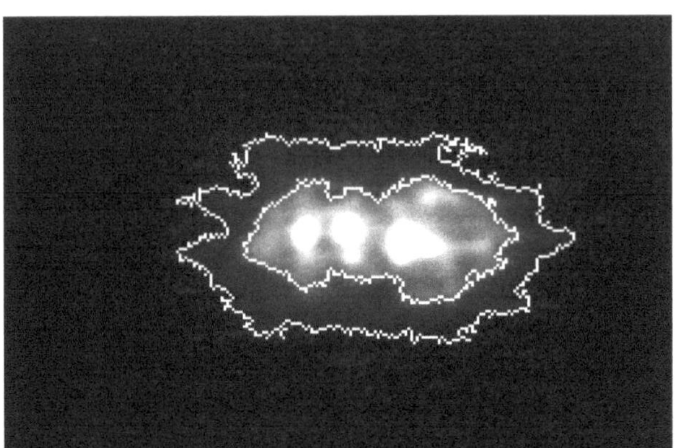

Figure 5.24 Contoured impact-damaged areas for a GFRP specimen impacted at $E = 8.3$ J.

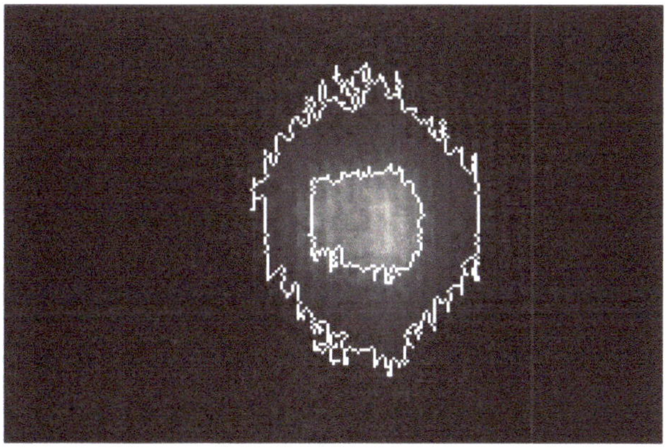

Figure 5.25 Contoured impact-damaged areas for a PPG specimen impacted at $E = 8.3$ J.

GFRP specimens impacted at $E = 2.8$ and 8.3 J, respectively, while Fig. 5.26 refers to a PPG specimen impacted at $E = 8.3$ J.

As can be seen by comparing Fig. 5.21 to Fig. 5.22, both A_{Di} and A_{De} increases with increasing impact energy. In particular, A_{Di} encloses only one hot spot (white dot) in Fig. 5.23 and three in Fig. 5.24. A_{De} displays a more-jagged outline because of the wedged delamination between fibres. By comparing Fig. 5.24 to Fig. 5.25 it is possible to see that the PPG specimen is characterized by more-circular and smooth-contoured A_{Di} and

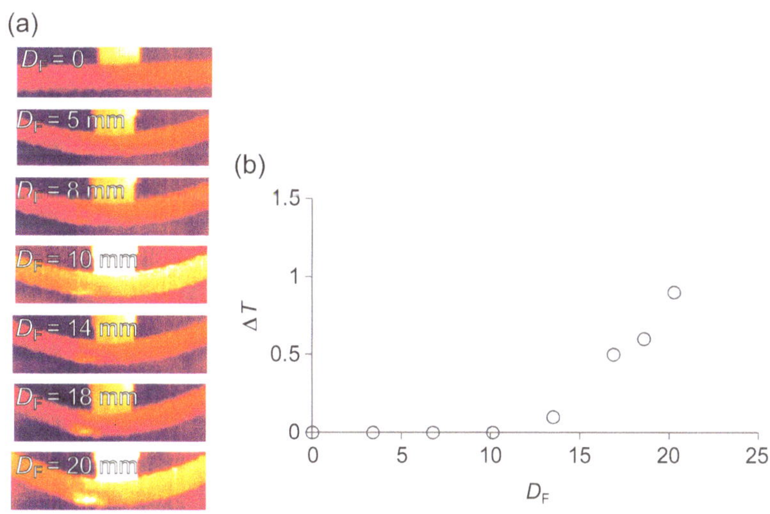

Figure 5.26 Thermal images and ΔT distribution along the specimen thickness for varying the displacement. (a) Some thermal images, (b) ΔT against D_F.

A_{De} areas. It is worth noting that again, as already described in Section 5.3.1, data are corrected for the alternate jump noise effect by using the unloaded reference area (see Fig. 5.12).

Finally, the extension of each zone can be measured by simply counting the pixels enclosed within the contour and considering the spatial resolution of the instrument (camera and lens) used [30]. This is not done herein because the scope is to describe a procedure and not to supply quantitative data and also because more tests are still necessary.

5.4.4 Main Outcomes

As the most important findings, a material under impact, depending on the impact energy E and on its intrinsic characteristics may experience:

- an elastic phase in which the impact energy is released without permanent material variations
- a plastic phase in which a part of the impact energy is absorbed with consequent permanent material modifications

The first phase happens for very low-impact energy, while the second phase becomes effective for a certain energy level, which depends on the material characteristics. The impact pushing force causes local bending of the surface; more specifically, on the side opposite to the impact the surface undergoes a convex curvature, which means local material expansion. From the thermal point of view, this means cooling down, or better, the thermoelastic effect already described in Section 5.2. For very low E (ie, elastic behaviour), once the force is removed, the surface attempts to recover its undisturbed position and its initial ambient temperature; no damage occurs. As E increases above the elastic limit, some damage occurs in the material, associated with heat dissipation and temperature increase; then, through the analysis of the local surface warming up, it is possible to acquire information about how the damage occurs.

This is possible because hot spots appearing over the viewed surface bear witness for loci of damage occurring in the material. By considering that during an impact event kinetic energy passes from the impactor nose to the target and that such energy is in part dissipated as heat, the detection of the heat generation loci helps for the comprehension of failure modes. In fact, any form of damage (delamination and/or fibres breakage) is accompanied by heat dissipation, which manifests itself through the appearance of hot spots/areas over the material surface. Indeed, it has already been demonstrated that the analysis of temperature variations during impact supplies important elements about damage onset and propagation as well [14,23]. In particular, by appropriate setting of testing parameters, it is possible to follow the evolution of thermoelastic/plastic phenomena.

In addition, from thermographic images it is possible to evaluate the extension of the impact-damaged and/or impact-affected area. This can be done after choice of a temperature difference threshold ΔT_T to be considered as a limit in the ΔT map to clearly identify a zone of a certain damage level.

5.5 QUASISTATIC BENDING TESTS

Quasistatic bending tests are performed owing to a three-points configuration with the load applied at the centre of the specimen, which is subjected to bending under the pushing force until collapse. In this case, an infrared imaging device may be used to visualize temperature variations over the specimen surface or over a lateral side (through its thickness). Any temperature rise (hot spot) is a symptom of formation of cracks. More specifically, it is possible through acquisition of images in time sequence to follow the evolution of cracks from their origin and successive grow up. This may be useful to get information about the material performance.

As an example, some thermal images, taken through the thickness, during bending at a speed of about 1 m/s are shown in Fig. 5.26 together with the distribution of ΔT values against the displacement D_F. For $D_F = 8$ mm, on the left of the punch, it is possible to recognize the presence of light tracts that account for local temperature rise due to formation of cracks. These tracts become lighter as the temperature increases and the cracks enlarge until collapse of the specimen for $D_F = 20$ mm. It is worth noting that these are only preliminary tests; some results have been presented for the sake of completeness.

5.6 CONCLUSIONS

This chapter has been completely devoted to the use of infrared thermography for monitoring online, during materials loading, time-dependent thermal phenomena, which may be exploited for the comprehension of the materials behaviour under load, which, in turn, are useful for assessing safe life criteria. Three types of loading have been addressed: cyclic bending, impact and quasistatic bending. Some results have been presented and discussed.

In particular, a general agreement between ΔT_a distribution and bending moment diagram has been shown and as such analogy is valid until the applied load is pure bending of a free-defect material. Any deviation induced by either a buried defect or different load application can be visualized by the infrared camera.

Some general indications about the measurement of the impact damaged/affected area have been given. A general criterion to be adopted for the choice of ΔT_T is necessary and is still not completely validated. The authors of this book are working in this direction and perhaps in the near future this gap could be filled.

For the sake of completeness, some preliminary results from quasistatic bending tests have been presented, and tests are under way involving different types of composites.

REFERENCES

[1] W. Thomson, On the thermoelastic, thermomagnetic and pyroelectric properties of matters, Philos. Mag. 5 (1978) 4–27.
[2] M.A. Biot, Thermoelasticity and irreversible thermodynamics, J. Appl. Phys. 27 (1956) 240–253.
[3] J.M. Dulieu-Barton, T.R. Emery, S. Quinn, P.R. Cunningham, A temperature correction methodology for quantitative thermoelastic stress analysis and damage assessment, Meas. Sci. Technol. 17 (2006) 1627–1637.
[4] U. Galietti, D. Modugno, L. Spagnolo, A novel signal processing method for TSA applications, Meas. Sci. Technol. 16 (2005) 2251–2260.
[5] P. Stanley, W.K. Chan, The application of thermoelastic stress analysis to composite materials, J. Strain Anal. 23 (1988) 137–142.
[6] T.R. Emery, J.M. Dulieu-Barton, J.S. Earl, P.R. Cunningham, A generalised approach to the calibration of orthotropic materials for thermoelastic stress analysis, Compos. Sci. Technol. 68 (2008) 743–752.
[7] C. Meola, G.M. Carlomagno, C. Bonavolontà, M. Valentino, Monitoring composites under bending tests with infrared thermography, Adv. Opt. Technol. 2012 (2012), 7 pp.
[8] C. Meola, G.M. Carlomagno, Infrared thermography to impact-driven thermal effects, Appl. Phys. A 96 (2009) 759–762.
[9] S. Boccardi, G.M. Carlomagno, C. Bonavolontà, M. Valentino, C. Meola, Infrared thermography to monitor Glare® under cyclic bending tests with correction of camera noise, in: Proc. QIRT 2014, Bordeaux, France, July 2014, pp. 7–11 paper 215.
[10] S. Boccardi, G.M. Carlomagno, C. Meola, G. Simeoli, P. Russo, Infrared thermography to evaluate thermoplastic composites under bending load, Compos. Struct. 134 (2015) 900–904.
[11] C. Meola, S. Boccardi, G.M. Carlomagno, Measurements of very small temperature variations with LWIR QWIP infrared camera, Infrared Phys. Technol. 72 (2015) 195–203.
[12] S. Abrate, Impact on Composite Structures, Cambridge University Press, Cambridge, 1998.
[13] R. Olsson, M.V. Donadon, B.G. Falzon, Delamination threshold load for dynamic impact on plates, Int. J. Solids Struct. 43 (2006) 3124–3141.

[14] C. Meola, G.M. Carlomagno, Impact damage in GFRP: new insights with infrared thermography, Compos. A 41 (12) (2010) 1839–1847.
[15] D.J. Elder, R.S. Thomson, M.Q. Nguyen, M.L. Scott, Review of delamination predictive methods for low speed impact of composite laminates, Compos. Struct. 66 (2004) 677–683.
[16] G.A. Schoeppner, S. Abrate, Delamination threshold loads for low velocity impact on composite laminates, Compos. A 31 (2000) 903–915.
[17] W.J. Cantwell, Geometrical effects in the low velocity impact response of GFRP, Compos. Sci. Technol. 67 (2007) 1900–1908.
[18] T. Mitrevski, I.H. Marshall, R. Thomson, R. Jones, B. Whittingham, The effect of impactor shape on the impact response of composite laminates, Compos. Struct. 67 (2005) 139–148.
[19] H.S. Huang, R. Talreja, Effects of void geometry on elastic properties of unidirectional fiber reinforced composites, Compos. Sci. Technol. 65 (2005) 1964–1981.
[20] P.O. Hagstrand, F. Bonjour, J.A.E. Månnson, The influence of void content on the structural flexural performance of unidirectional glass fibre reinforced polypropylene composites, Compos. A 36 (2005) 705–714.
[21] L. Liu, B.M. Zhang, D.F. Wang, Z.J. Wu, Effects of cure cycles on void content and mechanical properties of composite laminates, Compos. Struct. 73 (2006) 303–309.
[22] S. Abrate, Modeling of impacts on composite structures, Compos. Struct. 51 (2001) 129–138.
[23] C. Meola, G.M. Carlomagno, Infrared thermography to evaluate impact damage in glass/epoxy with manufacturing defects, Int. J. Impact Eng. 67 (2014) 1–11.
[24] C. Meola, G.M. Carlomagno, V. Lopresto, G. Caprino, Impact damage evaluation in composites with infrared thermography, 3rd European Conference for Aerospace Science EUCASS, Versailles (France) 6–9 July, 2009.
[25] G.M. Carlomagno, C. Meola, F. Ricci, Infrared thermography and piezoelectric patches for impact damage detection in composite structures, IWSHM 2011-International Workshop on Structural Health Monitoring 2011, Stanford University, Stanford, CA, September 13–15, 2011.
[26] C. Meola, G.M. Carlomagno, F. Ricci, V. Lopresto, G. Caprino, Investigation of impact damage in composites with infrared thermography, in: P. Mazal (Ed.), 6th NDT in Progress, Prague (Czech Republic) 10–12 October 2011, 2011, pp. 175–182.
[27] C. Meola, G.M. Carlomagno, Non-destructive evaluation (NDE) of aerospace composites: detecting impact damage using infrared (IR) thermography (inv. Chap.), in: Non-destructive Evaluation (NDE) of Polymer Matrix Composites, Woodhead Publishing Series in Composites Science and Engineering, vol. 43, Woodhead Publishing, Cambridge, UK, 2013. ISBN:0 85709 344 4, ISBN-13:978 0 85709 344 8 (Chapter 14).
[28] S. Boccardi, G.M. Carlomagno, C. Meola, P. Russo, G. Simeoli, Infrared thermography to evaluate impact damaging of thermoplastic composites, Proc. QIRT 2014, Bordeaux, France, July 2014. Paper 214 (2014) 7–11.
[29] C. Meola, S. Boccardi, G.M. Carlomagno, N.D. Boffa, E. Monaco, F. Ricci, Nondestructive evaluation of carbon fibre reinforced composites with infrared thermography and ultrasonics, Compos. Struct. 134 (2015) 845–853.
[30] S. Boccardi, G.M. Carlomagno, C. Meola, P. Russo, G. Simeoli, Monitoring impact damaging of thermoplastic composites, J. Phys. Conf. Ser. 658 (2015), http://dx.doi.org/10.1088/1742–6596/658/1/012005 (XXII AIVELA Annual Meeting) 012005.
[31] H.Y. Choi, H.Y.T. Wu, F.K. Chang, A new approach toward understanding damage mechanisms and mechanics of laminated composites due to low-velocity impact: Part II—analysis, J. Compos. Mater. 25 (1991) 1012–1038.

[32] H.Y. Choi, F.K. Chang, A model for predicting damage in graphite/epoxy laminated composites resulting from low-velocity point impact, J. Compos. Mater. 26 (1992) 2134–2169.
[33] Z. Aslan, R. Karakuzu, B. Okutan, The response of laminated composite plates under low-velocity impact loading, Compos. Struct. 59 (2003) 119–127.
[34] S. Abrate, Impact on laminated composite materials, Appl. Mech. Rev. 44 (1991) 155–190.
[35] T.J. Kang, C. Kim, Impact energy absorption mechanism of largely deformable composites with different reinforcing structures, Fibers Polym. 1 (2000) 45–54.
[36] C. Meola, G.M. Carlomagno, S. Boccardi, G. Simeoli, D. Acierno, P. Russo, Infrared thermography to monitor thermoplastic-matrix composites under load, Proc. 11th ECNDT, Prague, Czech Republic, October 2014, ISBN:978-80-214-5018-9 (2014) 6–10.

CONCLUDING REMARKS

This book was written with the intent to provide the reader with indications of the usefulness of infrared thermography within composite materials during fabrication and in service. Efforts have been made to make the matter easy to read and understand, even for nonexperts. Therefore, the theoretical part was minimized, trying to be essential without boring the reader with unnecessary digressions. Some notions on composite materials have been inserted with the purpose being not to teach anything on these materials but to facilitate the understanding of the utility of infrared thermography in their investigation. Also, a general description of the mostly in-use nondestructive testing techniques has been made while infrared thermography has been treated more deeply.

The attention was driven toward the use of infrared thermography with a twofold function of nondestructive technique and monitoring device. In fact, as shown in Chapter 4, infrared thermography can be used to detect either manufacturing defects, like fibre misalignments, voids, slag inclusions, or impact damage and/or degradation that have occurred in service. In Chapter 5 it has been demonstrated that infrared thermography may be a valuable means for online control of the behaviour of materials subjected to mechanical stress.

The obtained results prove that, through visualization of impact-induced thermal signatures, it is possible to getting information useful for the material characterization, specifically for identifying initiation and propagation of the impact damage. In particular, it is possible to circumscribe the whole area affected by the impact and to identify the bands of increasing damage. Also stressed, as already done in previous work, is the valuable and cost-effective use of an infrared imaging device to assess the extension of the impact-damaged area for design purposes. In fact, the use of an infrared imaging device allows for a rapid online appraisal, avoiding the waste of time in back-and-forth testing attempts, which is common practice in industrial enterprise to assess the performance under impact of new materials.

A lot has been done since the appearance on the market of the first infrared imaging device, but much still remains to be done. However, the technological progress with the continued release of new devices ever more sophisticated, more ergonomic, lighter etc. makes possible new applications, which also requires continued upgrading of procedures, data analysis methods etc. without end.

And perhaps, an infrared imaging device, if used in the most appropriate way, can help make the world safer!

INDEX

'*Note*: Page numbers followed by "f" indicate figures, "t" indicate tables.'

FSC
www.fsc.org
MIX
Papier | Fördert
gute Waldnutzung
FSC® C083411

Kolibri 360 GmbH
Ferdinand-Jühlke-Straße 7
99095 Erfurt, Deutschland
produktsicherheit@kolibri360.de